Pharmaceutical Packaging Technology

Second Edition

Pharmaceutical Packaging Technology

Second Edition

U. K. Jain
Professor & Principal
Bhopal Institute of Technology & Science-Pharmacy,
Bhojpur Road, Bangrasia, Bhopal (MP) 462 045 INDIA

D. C. Goupale
Asso. Professor
Bansal College of Pharmacy,
Bhopal Kokta, Anand Nagar,
Raisen Road, Bhopal (MP)
462 021 INDIA

S. Nayak
Professor & Principal
Bansal College of Pharmacy,
Bhopal Kokta, Anand Nagar,
Raisen Road, Bhopal (MP)
462 021 INDIA

Pharma Med PRESS
An Imprint of Pharma Book Syndicate

PharmaMed Press
An imprint of Pharma Book Syndicate
A unit of **BSP Books Pvt., Ltd.**
4-4-309/316, Giriraj Lane,
Sultan Bazar, Hyderabad - 500 095.

Published by :

PharmaMed Press

An imprint of Pharma Book Syndicate

A unit of BSP Books Pvt., Ltd.

4-4-309/316, Giriraj Lane, Sultan Bazar, Hyderabad - 500 095.
Phone: 040-23445605, 23445688; Fax: 91+40-23445611
e-mail: info@pharmamedpress.com

ISBN : 978-93-85433-26-9 (HB)

Preface to Second Edition

It is indeed a matter of great pride for us that the publication of second edition of our book came into being. The packaging of Pharmaceuticals is as much important as the medicine or drug, which is being packed. The quality to any medicine cannot only be imparted by imparting quality in its contents but it is of immense importance how this quality is being protected with continuously changing atmospheric conditions. Proper packaging is the only way to give desired shelf-life to pharmaceuticals.

It has been seen that a large number of medicines do not give the desired therapeutic effect due to faulty packaging. Packaging technology has become a separate subject in pharmaceutical sciences due to its vital applications in preserving therapeutic efficacy of drugs. Before starting the book writing on this subject, we noticed that the subject was less understood and it has been compiled with few chapters in some books. This inspired us to present detailed account in the form of a specialized book on the subject. Also the Packaging Technology has emerged as separate subject among specialized subjects of Pharmacy curriculum of almost all Universities.

Present book comprises of 16 chapters. The starting page of each chapter consists of chapter summary, which will help the teachers, students and other co-readers in revision of contents of whole chapter in only few minutes. Further the content of each chapter is arranged to facilitate the effortless flow of information in simple, motivating, and commonsensical approach. Detailed physico-chemical information on raw material e.g., glass, plastic, polymers, metals etc., used in Pharmaceutical packaging is presented to facilitate conceptual use of raw material for a particular type of medicinal preparation. Since the raw materials are common for packaging of other products, the scientific information may further be helpful to allied persons who deal with any kind of packaging.

A positive effort has been made to cover up all individual and specialized aspects of Pharmaceutical packaging sciences. More information is provided on Pharmaceutical packaging design, based on characteristic properties of materials used in packaging. Blisters and strips packaging are of growing importance and so a chapter is included on this topic which would be of great value to our readers. Parenteral and ophthalmic products need sterilized packaging; therefore a chapter on sterilization, covering unique features for sterilizing packaging materials, has been included. On demand of our readers from academics and industries, we have included a chapter on Blow-Fill-Seal (BFS) packaging technology in our upcoming second edition of this book.

Because of rising role of European Union as a world center for Pharmaceutical regulation and because of the significance of international harmonization of drug standards, a new chapter on regulatory aspects of Pharmaceutical packaging has been added. Chapters on quality control and stability testing of packages would explain the shelf-life concern aspects of the active pharmaceutical ingredient (API) packed.

It is our hope that this book will be broadly useful and, more generally, that it will facilitate the development of new dimensions of knowledge. Further, we hope that this book will be valuable not only to those with Pharmaceutical packaging persuasion but also to members of other disciplines who have concern in packaging material science.

-Authors

Contents

CHAPTER 1

Pharmaceutical Packaging and Design

Summary

1.1 Introduction

A product is not acceptably formulated except it is properly packaged, and in some cases the major part of the formulation process may be concerned with selecting the correct package for the product. **Packaging** is the science, art, and technology of enclosing or protecting products for distribution, storage, sale, and use. Packaging also refers to the process of design, evaluation, and fabrication of packages. Package labelling or labelling is any written, electronic, or graphic communications on the packaging or on a separate but associated label.

1.2 New Concept

Packaging of pharmaceutical products has come a long way in India and world scenario, but it still has a long way to go. Today's pharmaceutical industries look a lot colorful with blister packs and beautiful transparent bottles with bar-coding system of labelling. In other words packaging can be referred as the dress code of any pharmaceutical product. There are several reasons why packaging is becoming so crucial in present context.

1.3 Packaging Systems

Packaging may be looked at as several different types. For example a **transport package** or **distribution package** is the package form used to ship, store, and handle the product or inner packages. Some identify a **consumer package** as one that is directed toward a consumer or household. It is sometimes convenient to categorize packages systems by layer or function into following:

- Primary Packaging system
- Secondary Packaging system
- Tertiary Packaging system

1.4 Components of Packaging

A package consists of the container, closure, carton and box components. The container refers in which the product/medicine is placed and enclosed. Closure tightly packs the container. Carton is the outer covering, which gives secondary protection against mechanical and other environmental hazards and in the box multiples of products are packed.

1.5 Symbols used on Packages and Labels

Many types of symbols for package labelling are nationally and internationally standardized. For consumer packaging, symbols exist for product certifications, trademarks, proof of purchase, etc. Some requirements and symbols exist to communicate aspects of consumer use and safety. Recycling directions, Resin identification code (below), and package environmental claims have special codes and symbols. Bar codes (below), Universal Product Codes, and RFID labels are common to allow automated information management.

1.6 Package Development and Design Research

Package design and development are often thought of as an integral part of the new product development process. Alternatively, development of a package (or component) can be a separate process, but must be linked closely with the product to be packaged. Package design starts with the identification of all the requirements: structural design, marketing, shelf life, quality assurance, logistics, legal, regulatory, graphic design, end-use, environmental, etc. The design criteria, time targets, resources, and cost constraints need to be established and agreed upon.

1.1 Introduction

A product is not acceptably formulated except it is properly packaged, and in some cases the major part of the formulation process may be concerned with selecting the correct package for the product. **Packaging** is the science, art, and technology of enclosing or protecting products for distribution, storage, sale, and use. Packaging also refers to the process of design, evaluation, and fabrication of packages. Package labelling or labelling is any written, electronic, or graphic communications on the packaging or on a separate but associated label.

Packaging is greatly incorporated into our daily lives; we see it all around us, on everyday objects such as chocolate bars and potato chip packets. The main use for packaging is protection of the goods inside, but packaging also provides us with a recognizable logo, or packaging, we instantly know what the goods are inside. The last three decades had a constant development in awareness of pharmaceutical packaging. At least three factors seriously stimulated this concern, including

1. The desire to improve the efficacy of the dosage form through sophisticated dosage forms,
2. The increasing cost of packaging material
3. Consideration of Packaging drug regulations. The regulation intends to the importance of package so as to maintain the self-life of product and labelling descriptions on packages in achieving a physician's and customer's convenience regarding safe and efficacious use of pharmaceutical preparation.

It is very important to understand the key elements of regulations and techniques involved in the package systems.

1.2 New Concept

Packaging of pharmaceutical products has come a long way in India and world scenario, but it still has a long way to go. Today's pharmaceutical industries look a lot colorful with blister packs and beautiful transparent bottles with bar-coding system of labelling. In other words packaging can be referred as the dress code of any pharmaceutical product. There are several reasons why packaging is becoming so crucial in present context.

The Over the counter (OTC) and ethical environment is demanding better and more efficient packaging. From the market point of view many new concepts are being developed to improve overall usage of the product.

Some of the new concepts are emerging as per the need of manufacturer, market and socioeconomic conditions of consumer. For examples:

1. Packaging builds brand identity of product: Vicks has given the brand identity to the product.
2. Packaging way to differentiate the product with others: Electral packaging differentiates it with others.
3. Revolution in retail by display of product: attractive packaging for display especially in cosmetics and perfume products.
4. Patient convenience: some family planning products illustrate the way how it should be used
5. Innovation through pack design: Unit dose packs of herbal preparations like Hajmola candy.
6. Packaging that fulfills emotional needs: Various containers are made in toy shape to attract children e.g. Pack of Calcium Sandoz Tablets.

Packaging is as vital as formulation. If unsatisfactory packaging permits deterioration to set in before the patient uses it. Today it is clearly recognized that the product cannot exist without its package and a successful combination of the two is essential for a satisfactory shelf life.

Packaging and package labelling have several objectives:

Physical Protection: The objects enclosed, we are concerned here with the pharmaceutical dosage forms, in the package may require protection from, among other things, shock, vibration, compression, temperature, etc.

Barrier Protection: A barrier from oxygen, water vapor, dust, etc., is often required. Package permeability is a critical factor in design. Some packages contain desiccants or Oxygen absorbers to help extend shelf life. Modified atmospheres or controlled atmospheres are also maintained in some food packages. Keeping the contents clean, fresh, and safe for the intended shelf life is a primary function.

Containment or Agglomeration: Small dosages are typically grouped together in one package for reasons of efficient Packing and physical handling. For example, a single box of 100 ampoules of parenteral product requires less physical handling than 100 single ampoules. Liquids, powders, and flow able dosage needs containment.

Information transmission: Information on how to use, transports, recycle, or dispose of the package or product is often contained on the package or label. With pharmaceuticals, food products, medical, and chemical products, it is mandatory to quote this types of information by some regulatory agencies.

Marketing: The packaging and labels can be used by marketers for the purpose of encouraging potential buyers to purchase the product. Package design has been an important and constantly evolving phenomenon for decades of years. Marketing communications and graphic designs are applied to the surface of the package and (in many cases) the point of sale display.

Security: Packaging can play an important role in reducing the security risks of shipment. Packages can be made with improved tamper resistance to deter tampering and also can have tamper-evident features to help indicate tampering. Packages can be engineered to help reduce the risks of package pilferage: Some package constructions are more resistant to pilfeage and some have pilfer indicating seals. Packages may include authentication seals to help indicate that the package and contents are not counterfeit. Packages also can include anti-theft devices, such as dye-packs, RFID tags, or electronic article surveillance tags, that can be activated or detected by devices at exit points and require specialized tools to deactivate. Using packaging in this way is a means of loss prevention.

Convenience: Packages can have features, which add convenience in distribution, handling, display, sale, opening, reclosing, use, and reuse.

Portion Control: Single serving or single dosage packaging has a precise amount of contents to control usage. Bulk commodities (such as salt) can be divided into packages that are a more suitable size for individual households. It is also an aid for the control of inventory: selling sealed one-liter-bottles of milk, rather than having people bring their own bottles to fill themselves.

1.3 Packaging Systems

Packaging may be looked at as several different types. For example a **transport package** or **distribution package** is the package form used to ship, store, and handle the product or inner packages. Some identify a **consumer package** as one that is directed toward a consumer or household.

There are primarily two different types of packaging material under the present scope. These include primary packaging material, secondary packaging material and accessories (Table 1.1). The primary consists of those packaging components that from the part of the pack, which contain the product and are in direct contact of the product e.g. bottle, cap, liner etc. The main function of the primary pack is to contain or restrict the product and to provide protection to the product. All other packaging materials used external to the primary pack is known as the secondary packaging materials. These provide physical protection to ensure safe warehousing and mechanical protection required in shipment and transport. These include cartons, corrugated boxes, shipment container, etc. There may be some associated components in packaging, e.g. dosing dropper, calibrated spoon, etc.

Table 1.1 Primary, secondary, packaging material and accessories.

Primary	Secondary	Accessories
Glass bottles and jars	Cartons (paperboard boxes)	Pharmaceutical closures
Plastic bottles	Prescription dispensing containers	Dispensing closures
Strip packs	Corrugated boxes	Rubber stoppers
Blister packs	Paper drums	Paper, foil and laminated lids
Pouches	Shipping containers	Leaflets
Ointment tubes	Injection trays	Labels
Vials and ampoules		Shrink wrap and bands
IV containers		Wrappers
Pre-filled syringes		Inner seals
Aerosol containers		Gum tapes
Pre-filled inhalers		BOPP tapes
Paperboard containers		

It is sometimes convenient to categorize packages systems by layer or function:

1. Primary Packaging system
2. Secondary Packaging system
3. Tertiary Packaging system

1. **Primary packaging** is the material that first envelops the product and holds it; that is those package components and subcomponents that actually come into contact with the product, or those that may have a direct effect on the product shelf life. This usually is the

smallest unit of distribution or use. The examples of primary packaging are:-

Aerosol spray can	Bags-In-Boxes
Beverage can	Wine box
Bottles	Blister packs
Cushioning	Envelopes
Plastic bags	Plastic bottles
Skin pack	Tin can
Wrappers	

Primary packaging material should not:

1. have adverse effect on product due to chemical reaction, leaching, absorption or adsorption, particulate contamination,
2. be adversely affected by product,
3. be influenced by adverse manufacturing condition (e.g. sterilization, freezing).

2. **Secondary packaging** is outside the primary packaging – perhaps used to group primary packages together. Examples are-

Boxes

Cartons

Shrink-wrap

Secondary packaging component function

- Protection from excessive
  - ✓ transmission of reactive gases
  - ✓ moisture
  - ✓ light
  - ✓ microbes
- Protection to flexible containers
- Protection from rough handling during transportation

3. **Tertiary packaging** is used for bulk handling and shipping.

Barrel	Crate
Container	Edge protector
Intermediate bulk container	Pallets
Slip Sheet	Stretch wrap

Flexible intermediate bulk container, Big bag, "Bulk Bags", or "Super Sacks"

These broad categories can be somewhat arbitrary. For example, depending on the use, a shrink-wrap can be primary packaging when applied directly to the product, secondary packaging when combining smaller packages, and a tertiary packaging on some distribution packs.

1.4 Components of Packaging

A package consists of following components:

1. ***Container:*** The container refers in which the product/medicine is placed and enclosed. It remains in direct contact with the drug.
2. ***Closure:*** Closure tightly packs the container to exclude oxygen, carbon dioxide, moisture, and microorganisms and prevents the loss of water and other volatile substances form the product.
3. ***Carton/outer:*** It is the outer covering, which gives secondary protection against mechanical and other environmental hazards and also serves for the display of written information. The cartons are made up of cardboard, molded wood pulp and expanded polystyrene.
4. ***Box:*** In the box multiples of products are packed. The box provides primary defense against external hazards and have shock absorbing features. The boxes are made up of thick cardboard and wood.

1.5 Symbols used on Packages and Labels

Many types of symbols for package labelling are nationally and internationally standardized. For consumer packaging, symbols exist for product certifications, trademarks, proof of purchase, etc. Some requirements and symbols exist to communicate aspects of consumer use and safety. Recycling directions, Resin identification code (below), and package environmental claims have special codes and symbols. Bar codes (below), Universal Product Codes, and RFID labels are common to allow automated information management.

The symbols used in the Resin identification code consist of arrows that cycle clockwise to form a rounded triangle and enclosing a number, often with an acronym representing the plastic below the triangle. When the number is omitted, the symbol is known as the universal "Recycling Symbol", indicating generic recyclable materials. In this case, other text and labels are used to indicate the material(s) used. Where as the number inside the recycling symbol indicates about the number of recycling process applied on plastic material.

Shipments of hazardous materials or dangerous goods have special information and symbols as required by UN countries, and specific carrier requirements. Two examples are below:

With transport packages, standardized symbols are also used to aid in handling. Some common ones are shown below while others are listed in ASTM D5445 "Standard Practice for Pictorial Markings for Handling of Goods" and ISO 780 "Pictorial marking for handling of goods".

Fragile

Use no hand hooks

This way up

Keep away from sunlight

Keep away from water

Centre of gravity

Clamp as indicated

Do not clamp as indicated

1.6 Package Development and Design Research

Package design and development are often thought of as an integral part of the new product development process. Alternatively, development of a package (or component) can be a separate process, but must be linked closely with the product to be packaged. Package design starts with the identification of all the requirements: structural design, marketing, shelf life, quality assurance, logistics, legal, regulatory, graphic design, end-use, environmental, etc. The design criteria, time targets, resources, and cost constraints need to be established and agreed upon.

Transport packaging needs to be matched to its logistics system. Packages designed for controlled shipments of uniform pallet loads may not be suited to mixed shipments with express carriers.

An example of how package design is affected by other factors is the relationship to logistics. When the distribution system includes individual shipments by a small parcel carrier, the sorting, handling, and mixed stacking make severe demands on the strength and protective ability of the transport package. If the logistics system is for uniform pallet loads that are unitized, the structural design of the package can be designed to those specific needs: vertical stacking, perhaps for a longer time frame. A package designed for one mode of shipment may not be suited for another.

Sometimes the objectives of package development seem contradictory. For example, packaging for an over-the-counter drug might require **tamper resistance** and **child resistant** features: These intentionally make the package difficult to open. The intended consumer, however, might be handicapped or elderly and be unable to readily open the package.

Package design may take place within a company or with various degrees of external packaging engineering: contract engineers, consultants, vendor evaluations, independent laboratories, contract packagers, total outsourcing, etc. Some sort of formal Project planning and Project Management methodology is required for all but the simplest package design and development programs.

Sustainability involves responsible use of energy and resources. The package development process may involve a life cycle assessment which considers the material and energy inputs and outputs to the package, the packaged product (contents), the packaging process, the logistics system, waste management, etc. It is necessary to know the relevant regulatory requirements.

The traditional "three R's" of reduce; reuse and recycle have been expanded. All levels of the waste hierarchy may be considered in product and package development.

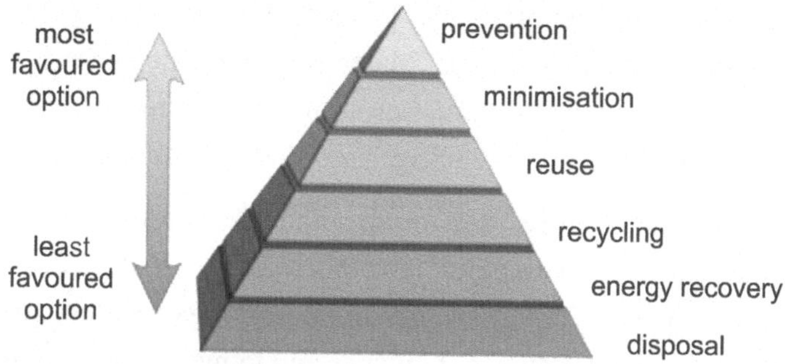

Fig.1.1 Favorable considerations during package development.

The waste hierarchy

- ***Prevention:*** Waste prevention is a primary goal. Packaging should be used only where needed. Proper packaging can also help prevent waste. Packaging plays an important part in preventing loss or damage to the packaged-product (contents). Usually, the energy content and material usage of the product being packaged are much greater than that of the package. A vital function of the package is to protect the product for its intended use: if the product is damaged or degraded, its entire energy and material content may be lost.
- ***Minimization:*** (also ''source reduction'') The mass and volume of packaging (per unit of contents) can be measured and used as one of the criteria to minimize during the package design process. Usually "reduced" packaging also helps minimize costs. Packaging engineers continue to work toward reduced packaging.
- ***Reuse:*** The reuse of a package or component for other purposes is encouraged. Returnable packaging has long been useful (and economically viable) for closed loop logistics systems. Inspection, cleaning, repair and recouperage are often needed.
- ***Recycling:*** Recycling is the reprocessing of materials (pre- and post-consumer) into new products. Emphasis is focused on recycling the largest primary components of a package: steel, aluminum, papers, plastics, etc. Small components can be chosen which are not difficult to separate and do not contaminate recycling operations.

- ***Energy recovery:*** Waste-to-energy and Refuse-derived fuel in approved facilities are able to make use of the heat available from the packaging components.
- ***Disposal:*** Incineration and placement in a sanitary landfill are needed for some materials. Material content should be checked for potential hazards to emissions and ash from incineration and leachate from landfill. Packages should not be littered.

Packaging, unit recently, was an after thought considered merely as a final step in manufacturing. But now packaging is thought of early in product development. The companies are using it as medium to promote their products, increase patient compliance and to meet recent regulations.

Pharmaceutical packaging has become an integral part of drug delivery technology. It is designed to provide protection against climatic, biological, physical and chemical hazards that the product may be exposed to during the product shelf – life. WHO guidelines define packaging as a process that a bulk material must undergo to become a finished product. The properties and attributes of the product should be as specified by the manufacturer and required by the user.

The pharmaceutical packaging can be defined as means for providing protection, presentation, identification, information, convenience, compliance and compatible unit, which maintain the integrity and stability of the product (Table 1.2). It should be economical and disposable. Package provides protection against climatic conditions (temperature, moisture, atmospheric gases, pressure, particulate matter and light), microbes (bacteria, fungi, yeast etc.) mechanical stress (during packaging operations, transportation, accidental drops and carriage), chemicals and misuse. Presentation helps in building the image of a product. Packaging helps in identification of the product and various desired information are displayed as per the label requirement of the product. It should be convenient to use or to administer the product e.g. eye drops, collapsible tubes, metered dose inhalers (MDI). It should be compatible with product and should neither contaminate the product nor be adversely affected by product. After exhaustion of the product, it should be easily disposed (recyclable or biodegradable). It should also take care of pilferage and adulteration risks.

Table 1.2 Role of packaging

- Protection
  - ✓ Against light
  - ✓ Against reactive gases
  - ✓ Against moisture
  - ✓ Against microbes
  - ✓ Against physical damage
  - ✓ Against pilferage and adulteration
- Presentation
- Identification
- Information
- Compatible
- Convenience

The packaging process includes filling and assembling, sterilization in the final container, if applicable, and placing labels on the container and storage at the manufacturing and shipping sites.

Packs are known to affect the shelf-life of all the pharmaceutical products including ethical, semi-ethical and proprietary. Transparent packaging is specially being encouraged to replace the opaque packaging as it allows streamline quality control, simplification of labelling, extended shelf-life and even enhanced product aesthetics. Additionally, the packaging should be economical and should provide information regarding the product and should also be compliant with the regulatory requirements.

Packaging design takes into account the needs of the product, manufacturing and the distribution system. The fundamental functions of packaging are to provide protection thereby not to allow leak, diffusion and permeation of the product; to be strong enough to hold the contents during normal handling and storage; not to be altered by the ingredients of the formulation in its final form. Some of the definitions regarding nature of the pack given in USP XXVII are listed below.

Definition (USP XXVII)

Well closed container: A well-closed container protects the contents from extraneous solids and from loss of the article under the ordinary or customary conditions of handing, shipment, storage and distribution.

Tight container: A tight container protects the contents from contamination by extraneous liquids, solids, or vapors from loss of the article, and from efflorescence, deliquescence or evaporation under the ordinary or customary conditions of handing, shipment, storage and distribution and is capable of tight re – closure.

Hermetic container: A hermetic container is impervious to air or any other gas under the ordinary or customary conditions of handing, shipment, storage, and distribution.

Light resistant container: A light resistant container protects the contents from the effects of light by virtue of the specific properties of the material of which it is composed, including any coating applied to it. Alternatively, a clear and colorless or a translucent container may be made light – resistant by means of an opaque covering, in which case the label of the container bears a statement that opaque covering is needed until the contents are to be used or administered.

Tamper-evident container: The container or individual carton of a sterile article intended for ophthalmic or otic use, except where extemporaneously compounded for immediate dispensing on prescription, shall be so sealed that the contents cannot be used without obvious destruction of the seal.

Innovations in packaging have been able to keep pace with the advances in product development through the application of newer materials, novel designs and latest packaging equipment, Newer polymers in plastics, newer low permeability films and foils, improvement in glass (e.g. coated glass) provide novel properties to the packaging. Just over half of medicines (51%) are taken orally as tablets and capsules that are packed in blister/ strip packs or fed into plastic bottles. Powder, pastilles and liquids also make up part of the oral medicine intake. other medicines include parenterals (29%), inhalation (17%) and transdermal (3%) preparations. Dosage forms are available in a variety of different shapes and sizes (e.g. tablets). These changes in dosage forms have an impact on packaging with an increasing need to provide tailored, individual packaging solution to maintain the quality, aesthetics and effectiveness of the dosage form. Need for specialized packages including child-resistant, senior-friendly have further increased expectations from the packaging industry.

Unit Dose and Device Packaging

Apart from primary and secondary packaging, two types of special packaging are currently in use. These include, unit dose packaging and device packaging. Unit dose packaging is defined as a single unit container for administration as a single dose, direct from the container. In this a single item or a specific quantity (dose) is enclosed within a disposable pack. These dosage units guarantee safer medication, are more practical, improve patient compliance and are more useful for less stable products. Device packaging comprises of packaging along with an administration device. This is user friendly and improves patient compliance. This allows an easier administration by means of devices such as pre-filled syringes, droppers, transdermal delivery systems, pumps and aerosol sprays. These devices ensure that the product is administered correctly and in the right amount. There are a number of packaging materials used for primary as well as secondary packaging. (Table 1.3).

Table 1.3 Type of raw materials used for pharmaceutical packaging

Type	Use
Glass	Bottles, Vilas, Ampoules, Syringes, Cartridges, IV Containers, Aerosol Container.
Plastics	Bottles, Syringes, IV containers, Tubes, Bags, Laminates, Pouches, Lids, Tapes, Aerosol, Container (dip tube, gasket housing button, stems).
Rubbers	Closures, Vial wrappers, Caps, Plungers.
Paper/ card boards Corrugated	Labels, Inserts, Display units, Pouches, Laminates, Cartons, boxes, Foils, Gum tapes, Paper drums.
Metals	Collapsible tubes, Foils, Needles, Aerosol container, Cans.

The choice of packaging material will depend upon:

(a) The dosage form desired.
(b) The degree of protection required.
(c) Compatibility with the dosage form,
(d) Presentation and aesthetics,
(e) Customer convenience e.g. size, weight of dosage form,
(f) Filling method,
(g) Sterilization method to be employed, and cost,

Apart from packaging material, the selection of pack is important. While selecting the package for a medication one must realize that the main object is to deliver a drug to a specific site of effective activity in the patient and the packaging must contribute to this end. The product/pack couple is designed as one unit to achieve the above objective. The choice of packaging for a specific pharmaceutical product is dependent upon:

(a) The nature of product and its compatibility with the material;

(b) The type of patient – whether child, elderly, adult;

(c) The type of dose – granules, tablets, ointment;

(d) Method and site of administration – dispensing device, etc.;

(e) Method of distribution – through hospital, pharmacy, retailer;

(f) Capacity of the packaging needed – small bulk for pharmacies, OPD; and

(g) Required shelf – life and likely sales area.

CHAPTER 2

Packaging Materials

Summary

2.1 Introduction

The materials selected for the packaging must have the following properties:

- Mechanical properties
- Physico-chemical properties
- Biological properties
- Economical aspect
- Pharmaceutical Properties

2.2 Materials for Construction of Containers

A variety of packaging materials are used as

- Primary packaging materials
- Secondary packaging materials
- Tertiary packaging materials

Primary packaging materials: included plastics, glass, metals, films, foils and laminates.

Plastics:

- Polyethylene (PE)
- Low density polyethylene (LDPE)
- High density polyethylene (HDPE)
- Polypropylene (PP)
- polystyrene (PS)
- Polycarbonate (PC)
- Polyvinyl Chloride (PVC)
- Nylon (Polyamide) (PA)
- Polyacrylonitrile polymers
- Polyethylene terephthalate(PET)
- Cellulose acetate
- Ethylene vinyl alcohol (EVOH)
- Polychlorofluoroethylene (PCTFE)
- Polyvinylidene Chloride (OvdC)
- Styrene acrylonitrile (SAN)
- Polytetrafluoroethylene (PTFE)
- Acrylonitrile butadiene styreme (ABS)

Glass:

Type I	Highly resistant borosilicate glass
Type II	Treated soda – lime glass
Type III	Soda – lime glass
Type NP	General purpose soda

Metals: Some metals currently in used in the pharmaceutical packaging include tin plate, tin – free steel, aluminium, alloy of aluminium and stainless steel.

Film, foils and laminates: Film, foils and laminates perform different roles, such as supportive, barrier, heat seal and decorative. Paper is usually used as a supportive ply, which can be readily printed to give a decorative appeal.

Secondary packaging materials: Paper and board are broadly cellulose/ natural fiber materials. They find more diverse applications when compared with primary packaging materials. Cartons/corrugated board made from these materials is relatively recent invention.

Tertiary packaging materials: It includes barrel, crate container, edge protector, Intermediate bulk container, pallets, slip-sheet, stretch wrap, flexible intermediate bulk container, big bag, "bulk bags", or "super sacks".

2.1 Introduction

It is very important while selecting a package that the selected package should effectively preserves the integrity of the pharmaceutical product. The packaging materials must behave in a 'neutral' way and contribute to the stability of the product against ecological factors causing deterioration. The choice of package hence starts with the study of physicochemical characteristics, protective requirements and marketing necessities of the product to be packed. The materials selected for the packaging must have the following properties:

- Mechanical properties
    - Container should be adequately strong enough to withstand the shock during handling, filling, closing, storage and transport.
    - It can withstand heat during sterilization.
    - It should be impermeable, to avoid any loss of product, or contamination by liquids, gases, vapors or microorganisms.
- Physico-chemical properties
    1. The container should not absorb substances from the product
    2. It should protect light sensitive drug.
    3. It should be non-reactive with content of the preparation.
    4. It should not impart its own color, taste and odour to the preparation.
    5. Container and closure should not react with each other and with the preparation.
    6. Closure must be easy to remove and replace.
    7. The materials used in container should be F.D.A. approved.
- Biological properties
    1. The container should not produce any toxic effect.
    2. It should not affect the therapeutic value of the product.
    3. It should have a suitable life, which should be comparable with the stability period of the product.
    4. It should be able to withstand attack by insect and should not support mould growth.
- Economical aspect
    1. Cost wise the container and closure should be reasonable.
    2. The cost of container should be proportionate with the cost of the product.

- Pharmaceutical Properties

  Followings Pharmaceutical properties should be taken into consideration while selecting the materials for Pharmaceutical packaging.

  1. Dosage form of the product – Solid, semisolid, liquid or gas.
  2. The route of administration of preparation: Oral, parenteral, external.
  3. The stability of the products towards: Light, oxygen, moisture, carbon dioxide, and trace metals.

2.2 Materials for Construction of Containers

I. Primary Packaging Material

All the above-mentioned aspects for the selection of packaging material have their own significance; even though, protection including mechanical, climatic and biological protection is always the most serious factor as it determines the shelf life of a product. A variety of packaging materials used as primary packaging materials, included plastics, glass, metals, films, foils and laminates. The brief discussion of these packaging materials is presented below:

Plastics

A plastic packaging material is supposed to be robust, strong, light, aesthetic, pilfer-proof, protective, easy to carry, convenient to open, use or administer, close, dispose and tamper evident. They should be competent of being used in hi-speed mechanical processing. It also requires to be created simply at economical cost. Some time formulation package needs to be child-resistant and sterilizable. Due to availability of range of polymers, plastic can satisfy the requirements stated above. These can be used as homo-or co-polymer or in combination with aluminium and board. The advantages and limitations presented by plastics as packaging material are listed below:

Advantages

1. Plastic materials have very good mechanical strength
2. They are non-breakable

3. Range of plastic material available, which may satisfy requirement of the pharmaceutical packaging.
4. They reduce the freight cost due to their less weight.
5. Plastic containers are obtainable in multipurpose design.
6. Plastics are poor conductor of heat.

Disadvantages

1. Plastics are permeable by vapour.
2. The presence of many additives may brings about the Drug-plastic interaction.
3. The majority of plastics are sensitive to heat.

Plastics include a group of material of natural or synthetic origin consisting mainly of polymers of high melting point, capable of being molded into variety of shapes or forms by utilizing heat or pressure. These consist of molecules called the monomers. Monomers go through a process of polymerization forming a plastic. The polymerization process may involve various chemicals, which help in polymerization such as initiators, solvents, catalysts, accelerators, etc. Some examples of additives present in plastics are listed below:

Additives, processing aids and residues that may be present in plastic containers

- Acceleration
- Antiblocking agent
- Antioxidants
- Antislip additives
- Anti static agents
- Catalyst
- Colorant
- Emulsifier
- Extenders
- Fillers
- Flame retardants
- Lubricants
- Light excluders
- Modifiers
- Mould release agents
- Opacifiers
- Plasticizers
- Stabilizers
- UV absorbers

Plastics may be amorphous or crystalline in nature. Amorphous plastics are more permeable to gases and moisture but are less inert. They offer good clarity, transparency, and hardness with possible brittleness. Crystalline plastics have low permeability to gases and moisture and are more inert. They are opaque or translucent and more flexible. The polymers used in pharmaceutical packaging are listed below:

Polymers as packaging materials

• Polyethylene (PE)	• Polyethylene terephthalate(PET)
• Low density polyethylene (LDPE)	• Cellulose acetate
• High density polyethylene (HDPE)	• Ethylene vinyl alcohol (EVOH)
• Polypropylene (PP) (PCTFE)	• Polychlorofluoroethylene
• polystyrene (PS)	• Polyvinylidene Chloride (OvdC)
• Polycarbonate (PC)	• Styrene acrylonitrile (SAN)
• Polyvinyl Chloride (PVC)	• Polytetrafluoroethylene (PTFE)
• Nylon (Polyamide) (PA)	• Polyacrylonitrile polymers
• Acrylonitrile butadiene styrene (ABS)	

Polymers used in plastics can be divided into two categories:

1. ***Thermosetting polymers:*** A polymerization process involving a curing or vulcanization stage during which the material becomes "set" to a permanent state by heat or pressure produces thermoset ploymers. These comprise of phenolics, melamine, urea, epoxies, certain polyesters and polyurethanes.
2. ***Thermoplastics polymers:*** Thermoplastic polymers are heat-softening materials, which can be repeatedly heated, made mobile and then reset to a solid state by cooling. These comprise of more number of polymers e.g. polyethylene, polyvinyl chloride, polystyrene, polypropylene nylon, polyester, polyvinylidene chloride, polycarbonate.

Further polymers may be divided into categories such as homo-polymer, co-polymers, terpolymers depending upon whether ploymerization involves one, two or more types of monomers respectively.

The increasing demand of plastics in the pharmaceutical industry and their vast applications may be attributed to the ease of formation of plastic containers, light weight, availability of high quality plastic container, resistance to breakage and liberty to design the container in several shape and size. A great advantage of working with plastic material is the almost never-ending potential of creating novel designs. However, permeation / leaching of components from plastic containers into product may create problems. The problem can be set on using high quality grade polymers for the preparation of plastic pharmaceutical containers.

Mainly plastic containers used in pharmaceutical products are made up of the following polymers: low, medium and high density polyethylene (PE), plasticized and unplasticized polyvinyl chloride (PVC), homopolymer and copolymer of polypropylene (PP) and polystyrene (PS). Plastic containers also include polymethyl methacrylates (PMM), polytrifluoroethylene, nylon (PA), acrylonitrile betadiene styrene (ABS), styrene acrylonitrile (SAN), polycarbonate (PC), etc.

Plastic pharmaceutical containers comprise plastic bags for parenteral, plastic bottles for liquids, plastic trays for tablets and capsules, tubes for creams and ointments, jars, boxes of powder, etc. A variety of different plastics and their blends are used for this purpose, including the following main polymers.

Polyethylene (PE)

Polyethylene is obtainable in three different grades with low (LDPE), medium (MDPE) and high (HDPE) density (according to the density of the PE ranging from 0.91 to 0.96) As the density of PE raise both physical and chemical properties vary. The raise in polymer density makes the material more rigid, stronger, less impact resistant, less translucent, less flexible with an increase in melting point. The clarity and translucency also depend upon the density of PE. The low, medium and high density PE have oxygen transmission of 500, 535-250 and 185, respectively. Anti oxidants and antistatic additives are used in bottle grades of polyethylene.

High-density polyethylene (HDPE) is the most crystalline material and is most broadly used for containers by the pharmaceutical industry. It offers a good barrier against moisture but a relatively poor one against oxygen and other gases. Most of the solvents do not react with PE and it unaffected by strong acids and alkalis. HDPE generates most applications in standard rigid containers for solid oral medications.

Polyethylene is a material of choice along with polypropylene (PP) as a polymeric material used for on-line blow-fill-seal (BFS) technology. These are inert and give a good balance of properties for BFS technology including ease of forming, opening and handling of the finished containers.

Linear low-density polyethylene (LLDPE) is a relatively newer material, which is getting a lot of importance as a film forming material. LLDPE is being used for plastic tubes and as a seal ply to provide high mechanical strength.

Polypropylene (PP)

Polypropylene is somewhat similar to HDPE, has a remarkably low density and hence is capable to produce more finished products than a similar weight of any other thermoplastic resin. It also has an excellent resistance to almost all types of chemicals, including acids and most organic materials. It does not break under any condition and gives almost complete barrier against gas and vapor. Its high melting point makes it suitable for sterilizable products. It has a drawback of being brittle at low temperature. Though it has good impact strength and impact durability, improvement in low temperature impact strength is required in most case. This improvement is generally achieved through the use of an elastomer.

Polystyrene (PS)

Polystyrene is a clear, rigid, hard material with good tensile strength but is the most brittle plastic when dropped. It is resistant to mineral oils, water but is soluble in organic solvents. It is somewhat permeable to moisture and is usually not an appropriate packaging material for pharmaceutical products.

Polycarbonate (PC)

Polycarbonate is having a good impact resistance with excellent dimensional stability. It is heat resistant and has a low water absorption capacity. It is used to make membrane filters, reusable bottles and sterilizable pharmaceutical packaging.

Polyvinyl chloride (PVC)

Polyvinyl chloride (PVC), amongst the vinyl polymers including polyvinyl chloride (PVC), polyvinylidene chloride (PVdC), polyvinyl acetate (PVAc) and polyvinyl alcohol (PVA); offers utmost applications. It is an inexpensive, tough, clear material that is relatively easily processed. It is available as un-plasticised PVC (UPVC), plasticised PVC or as an impact modified PVC. It may show some deterioration upon ageing; therefore, stabilizers are invariably added into these to improve stability. The amount of plasticizer in PVC may be greater than the amount of PVC; which may lead to reduced chemical resistance and increase gas/ moisture permeation. Plasticised PVC is soft and flexible.

Bottles of PVC with crystal clarity, rather good oxygen barrier, and greater stiffness can be produced. PVC, in its purest form, is crystal clear and stiff with poor impact resistance. It can be softened with the help of plasticizers. Stabilizers, modifiers, monomer residues, fillers, antioxidants, colorants can be incorporated into PVC to impart the preferred attributes. It can be plasticized to produce articles, which range in flexibility from soft to hard. It has a good resistance to chemicals and water permeation and an excellent elasticity.

Both PVC and HDPE occupy the largest quantity of raw materials employed in the production of pharmaceutical bottles. PVC is employed extensively in the production of semi-rigid and flexible bottles for low barrier liquid pharmaceuticals such as contact lens solutions and laxatives. Rigid PVC has been used for many years for providing protection to the pharmaceuticals.

PVdC has an good resistance to permeability by moisture and gases, very good chemical resistance with toughness, flexibility and heat sealability. PVdC, although existing as a film, is more usually found in aqueous dispersion or a solvent coating. polyvinyl acetate and copolymers find use as an adhesive and a heat sensitive base. Polyvinyl alcohol has excellent grease resistance and good gas barrier properties when dry. It has been used to produce water – soluble sachets.

Polyethylene terephthalate (PET)

The most usually used film for tape is produced by the reaction of dimethylterephthalic acid and ethylene glycol to give polyethylene (glycol) terephthalate. PET films are tough and highly tear-resistant. PET is used for preparation of stretch blown containers and also used pharmaceutically for small molding.

Advanced olefin copolymer (COC)

The polyolefins group includes ultra and very low-density polyethylene, low-density polyethylene (LDPE), linear low-density polyethylene (LLDPE), medium density polyethylene (MDPE) and high-density polyethylene (HDPE). It also includes ultra high molecular weight grades of HDPE. Advanced olefin copolymer, the cycloolefin copolymers (COC) is new, high barrier thermoforming plastic suitable for pharmaceutical and cosmetic packaging with excellent moisture barrier properties, excellent transparency and good rigidity. The use of high purity COC polymer for preparation of vials offers transparency as compared to glass. This polymer provides excellent barrier properties, high chemical resistance and breakage resistance. These attributes make this polymer ideal for primary packaging of pharmaceuticals. A outline of salient properties of different plastic used in pharmaceutical packaging is presented below:

Outline of salient properties of various plastic used in pharmaceutical packaging.

Plastic types Properties	PE	HDPE	PVC	PP	PS	PA	PC	Polymethyl methacrylate	Phenol Urea	Formal dehyde
Light sensitive	√	√								
Rigid	√	√								
Flexible	√									
Nontoxic	√		√							
Air	√	√	√	√			√			
Impermeable										
Moisture	√			√	√	√		√	√	√
Impermeable										
Autoclavable	√	√		√		√		√		
Electrostatic	√	√			√					
Oil resistant		√								
Translucent		√		√		√		√		
Transparent			√	√		√				
Heat-Resistant	√			√		√	√		√	√
Impact-Resistant						√	√	√		

The European pharmacopoeia provides a list of plastics that are permitted for use in pharmaceutical containers. These are:

- PVC for containers for human blood, blood components and aqueous solutions for IV infusion
- PVC for components used in blood transfusion
- Polyolefins
- LDPE for parenteral and ophthalmic preparations
- HDPE for parenteral preparations
- PP for parenteral preparations
- Ethylene – vinyl acetate copolymer (EVAc) for total parenteral nutritional (TPN) products
- Silicone oil as lubricant and
- Silicone elastomer for closures and tubing.

The lists of approved additives include antioxidants, stabilizers, plasticisers, lubricants, color and impact modifiers. For containers for oral and external preparations, anti – static and mould released agents can be used only. For choice of appropriate polymer material, the key aspects are that the drug is not absorbed; it does not migrate through the material, and does not yield any material is quantity sufficient to alter the stability or the toxicity of the product. Tests of compatibility include physical changes, permeation, pH change, effect of light and chemicals, and biological testing. Plastic containers of British Pharmacopoeia have been categorized into three different types:

(a) Plastic containers for aqueous solutions for parenteral infusion: The most commonly used polymers include polyethylene, polypropylene and PVC. They may be in shapes of bags or bottles. These must withstand the sterilization conditions to which the containers are subjected, be impermeable to microorganism and should be sufficiently transparent to allow visual inspection.

(b) Sterile plastic containers for blood and blood components.

(c) Empty sterile containers of plasticised PVC for blood components.

Plastics form a very significant part of the pharmaceutical packaging industry. There are several procedures and processes, which would never have been possible but for plastics. These include prostate devices, catheters, tracheotomy tubes, and blood collection containers, to name a few. These also form secondary packaging containers. Wide ranges of plastics are available for these purposes. A careful selection of the material is required with regard to the customer usage, manufacturability, sterility, barrier properties, etc. Depending upon the target material properties, likely material candidates are determined by comparing their properties with the property profile derived form the functional requirements. There are various material properties, which affect the functional performance of plastics, including mechanical, optical, and electrical properties. Electrical properties may cause hazard to electronic instruments and may attract dirt whereas the mechanical and optical properties have more relevance. The various mechanical properties tested in case of plastics include the following:

1. ***Impact strength:*** It provides a measure of the ability to withstand shock loading. The test material may get a blow from a swinging pendulum. The material will crack if the impact force exceeds the limit of elasticity of the material. Glass has such lower impact strength as compared to plastics.

2. ***Tensile strength:*** It provides a measure of the maximum force required to pull apart a specimen of material per unit area. Elongation measures the ability of the material to stretch.
3. ***Stiffness:*** It provides a measure of resistance of bending against a load.
4. ***Tear strength:*** It provides the measure of force required tearing a plastic and also the fore necessary to propagate the tear. While this property is undesirable in case of shipping sacks yet it is desirable in case of tear tapes.
5. ***Flex resistance:*** It provides a measure of the resistance to develop pinholes or fracture, when subjected to repeat flexing.
6. ***Coefficient of friction of slip:*** This property of plastic has more significance to form-fill-seal (FFS) operations. This depicts the ease of sliding of one material film over the other that is required in packaging machinery. Reverse of this is blocking, which gives the tendency of different films to stick together.
7. ***Fatigue resistance:*** It provides the ability of a plastic to withstand repeated short time stress/ deformation without cracking.

In addition to the above general tests, some other tests, which are specific to the product requirements e.g. behavior at low temperature, melting point, heat sealability, sterilizability, etc. are also performed. The important optical properties as in case of plastics include

1. ***Light transmission:*** It provides the measure of light transmission through a film.
2. ***Clarity:*** It provides an indication of degree of distortion of an object seen through the film of the material under test.
3. ***Haze:*** It provides a measure of haziness caused during product inspection through a film of the material.
4. ***Glass:*** It provides a measure of brightness on the surface.

All the plastic material are permeable to varying degree to moisture, oxygen, carbon dioxide, etc. There are also other disadvantages including possible extraction, interaction, adsorption, poor stability, etc. Moreover plastics also suffer from various negative features like:

1. ***Cavitation/panelling:*** Plastics are prone to distortion, partial collapse, swelling or dimpling.
2. ***Stress cracking:*** Low-density plastics are also prone to cracking under stress.

3. ***Crazing:*** This happens particularly in case of polystyrene (PS) and some other substance where a surface reticulation occurs, which may grow and lead to disintegration.
4. ***Poor printability:*** Plastics such as polyolefins need particular pretreatment before ink may be applied on them.
5. ***Poor impact resistance:*** Some plastics have a poor impact resistance e.g. polystyrene, PVC, etc.

Majority of these limitations can be overcome by using combination of plastics with additives, other plastic materials, films, foils, laminates, etc.

Glass

Glass is composed primarily of sand, soda-ash, lime-stone and cullet. Where sand represents for almost pure silica, soda-ash for sodium carbonate and lime-stone for calcium carbonate. Cullet is broken glass that is mixed with the batch, which acts as fusion agent for entire mixture.

Glass has served the pharmaceutical industry for a couple of centuries, till the advent of plastics after which there has been a competition on the use of one with respect to the other. However, recently the advantages of both the containers have been put together made in union in the form of plastic coated glass. Advanced glass containers are lightweight with a surface treatment. A number of glass containers used in pharmaceutical industry include ampoules, bottles, vials, syringes, and cartridges.

The many advantages that glass offers have enabled it to exist in the pharmaceutical packaging till date are presented in table below. Glass offers superior quality protection. It is economical, chemically inert impermeable with no diffusion or leakage (for good quality glass), strong and rigid, recyclable, hygienic, sterilizable, easily washable, resistant to high temperature, compatible and transparent. Another particular advantage that glass can offer is light protection. The advantages and limitations presented by glass as packaging material are listed below:

Advantages

1. Glass as packaging material is chemically inert
2. It impart no odour and taste to the product
3. It is non corrosive
4. It is strong and rigid
5. It is impermeable to vapours.
6. It is transparent and sparkle
7. Glass posses FDA approval
8. It does not undergo any change on aging

Disadvantages

1. Glass is fragile
2. It offers less pressure safety and impact resistance (aerosol containers can withstand pressure below 25 psig and use of less than 50% propellant is quite safe)
3. It Imparts no odour and taste to the product
4. we have to pay high freight cost due to its heavier weight

Several types of glass are available. These have been categorized as per USP into four different types. Type I glass is neutral glass and requires a higher working temperature and has a narrow working range thereby is difficult to process. These have a high resistance to chemicals. Type II is a soda glass, which has had a surface treatment by sulphuring or sulphating, forming a coating of sodium sulpahte, which neutralized the excessive surface alkalinity. Glasses are being given different treatments in order to improve their surface lubricity, increasing impact resistance and aesthetic appearance. Surface treated glass find wide application in pharmaceutical industry. Type III glass is the regular soda lime glass in which the containers are untreated and have an average chemical resistance. They have a limited alkalinity. Type NP glasses are the general – purpose soda lime glass and are supplied for non- parenteral product, intended for oral or topical use. Some specific features of all these 4 types of USP glass are presented below:

Specific features of glass (USP)

Type	General Description	Properties	Suitability
I	Highly resistant borosilicate glass	Alkalinity is removed by using boric oxide to neutralized the oxides of potassium and sodium.	Preparations for parenteral administration.
II	Treated soda – lime glass	Obtained by treating the hot surface of type III glass by sulphur dioxide/ammonium sulphate/ ammonium chloride	Preparations for parenteral administration
III	Soda – lime glass	It is an alkaline glass having high percentage of lime and soda and no boric oxide as compare to type I glass.	Not generally used for parenteral preparations until and unless indicated
NP	General purpose soda – lime glass	It has similar compositions to that of Type III glass but there is no guarantee of similar properties	Non parenteral articles e.g. oral or topical use

The British Pharmacopoeia categorizes the glass containers according to hydrolytic resistance into 4 types.

1. *Type I:* neutral glass for parenteral use for human blood and blood components.
2. *Type II:* high hydrolytic resistance glass suitable for acidic and neutral aqueous preparations for parenteral use with a pH lower than.
3. *Type III:* soda lime glass with moderate hydrolytic resistance suitable for non- aqueous preparations for parenteral use, powder for parenteral use and for non-parenteral preparations.
4. *Type IV:* soda lime silica glass with low hydrolytic resistance suitable for solid preparations that are not for parenteral use and for some liquids or semi – solid preparations that are not for parenteral use.

Glass is used for different containers ranging from ampoules, vials, cartridge tubes, disposable syringes and aerosol. Ampoules are generally made up of neutral glass. These were one of the first unit dose containers but are increasingly being replaced by cartridge tubes and pre – filled syringes. vials on the other hand are produced by soda glass, when used for tablets and capsules. Injection vials can still be obtained in either neutral or soda glass of occasionally in treated soda glass. Disposable glass syringes have been replaced by plastic syringes. Glass containers for aerosol purpose present risk of breakage as compared to the metal containers. Metal containers on the other side do not offer flexibility in design. Glass is being used for aerosol purpose making use of good bottle strength or an adequate bottle strength with an external coating of a flexible plastic (e.g. PVC). Uncoated bottles are used for keeping low pressure aerosols.

After the advent of plastic as a packaging material, it was once considered that the application of glass would come down drastically. However except for the breakage risk, glass the best product protection with distinct advantages. This is testified also with the fact that. FDA acceptance of glass rarely presents any problem. Lightweight coated glass with improved color range and strength seem to be very promising for pharmaceutical packaging in the future.

Films/ Foils and Laminates

Film, foils and laminates perform different roles, such as supportive, barrier, heat seal and decorative. Paper is usually used as a supportive ply, which can be readily printed to give a decorative appeal. Aluminium on the other hand is used for its barrier properties and for decorative appeal. Metallization is another process, which is relatively new, wherein particles of metal are laid down to a surface under vacuum to alter the surface properties of materials. However, this is not able to provide as good barrier properties as achieved by aluminium. Plastics as film or coating can be used for decoration, flexibility, barrier purpose, heat seability, transparency, and to protect the other piles within the lamination.

Films, foils or laminates are made up of a single layer, multi –layers or a combination of materials. Single ply materials include paper and plastics, which do not require an additional coating achieve heat seal or can be employed as a direct wrap (polyethylene and palsticized PVC).

Single ply materials: These including LDPE, LLDPE, or a mixture of blend involving combination of LDPE, MDPE, HDPE, EVAc (ethylene vinyl acetate), etc. find a wide application for the formation of bags, sacks, sachets, over – wraps, shrink wraps, etc. Deep freeze packs make use of LDPE, LDPE mixtures. Single ply plastics are also used in window cartons (cellulose acetate, polyester, regenerated cellulose) and in plastic cartons (PVC).

Shrink-wraps: shrink-wrap utilizes an oriented plastic which, when placed around an object, can be heated to a temperature where it returns to its original dimensions thereby forming a tight shrink-wrap around the object. LDPE, PP, PVC, EVAc have been used for this purpose.

Stretch wraps: These films must be elastic in nature, once in position must not relax and lose tension, LLDPE is the main material used for this purpose. LDPE, EVAc and unplasticised PVC films may be used in combination.

Cling films: These are an elastic – type material that undergoes stretch. PVC and PVdC were earlier used as cling films but recently LLDPE and LDPE mixture are being used for the purpose.

PVC is used in over wrap films, shrink sleeving, and thermoforming for blisters and for bubble packs. PVC films may be plasticized PVC films or Unplasticized PVC films. The former types are used for forming pillow pack as these are highly permeable is moisture. Unplasticized PVC films have low permeability to oxygen but moderately permeable to moisture. PVdC films are soft but very strong and are difficult to handle. They provide an excellent barrier to moisture and gases. polyvinylidene fluoride (PVdF) is more inert as compared to PVdC but is more expensive.

polyvinylidene fluoride (PVF) offers a high weather resistance but its again very expensive. polystyrene (PS) is used for blister – type packs. It is highly permeable to moisture and fairly permeable to gases.

Films

Regenerated cellulose film (RCF)

These films, though continuous, resemble paper in its properties, are moisture sensitive and not heat sealable, may be coated with nitrocellulose or PVdC on one or both the sides. Even after coating these films remain moisture sensitive and need storage under controlled conditions.

Films and coatings based on plastics

These films are from organic origin. They are continuous, clear, coloured or opaque films with a glossy surface. They are flexible, strong and sealable. They resist water and form effective barrier against water vapour depending upon their caliper, area, gradient, temperature, etc.

Special films

The polyolefins include LDPE, MDPE, HDPE, LLDPE etc, and are used for formation of bags and to form heat seal inner ply in laminations, sacks, shrink and stretch films. LDPE films are frequently used as a lamination ply to bond two materials together. HDPE films are employed in boil – in – the – bag applications.

PVdC films are soft cling films but are very strong and form excellent barrier properties. Polystyrene films lack barrier properties and may be used in blister – type packs for non – barrier usage. Aclar films are ploymonochlorotrifluoroethylene (PCTFE) films that offer high inertness and excellent barrier properties. Polytetrafluorethylene (PTFE) is used as a coating for closures. Polyvinylidene fluoride ((PVdF) and polyvinyl fluoride (PVF) are high barrier films but they are very expensive. Polyester (PET) films form good barrier against gases but only fair to moisture. It may be used as an outer ply to laminates.

Foils

Aluminium foils: An Aluminium foil is the material most widely used for packaging of tablets and capsules though it is the most expensive constituent of the laminate. Aluminium foils is made from the metal that has 99% purity and the gauge of the foil may vary from 0.006 to 0.040 mm. Aluminium foil offers excellent barrier properties to moisture, gases and light, gives the pack brightness and an attractive look. The foil provides hygienic, tamper evident pack, which is odorless, tasteless and non – toxic. It provides a convenient, safe and versatile packaging for tablets, capsules, liquids and powders.

Aluminium foil is strong and can be laminated. It can form push – through lids of preformed trays for tablets. When used as collapsible tube these do not draw air into them (forming dead ends), which may contaminate the product.

Aluminium foil is particularly suitable for blister packs. The tray of the blister may be made of PVC/ PVdC while the lidding is done with the help of aluminium foil. Aluminium foil can be made into alloy, which can alter its properties like brittleness, toughness and ductility. Aluminium foil is also used for forming strips packs. Strip packs provide economical high barrier performance pack and can be filled at high speed on modern, sterile filling lines. In case of sachets and pouches aluminium packs are robust and keep the product in good condition. These can withstand rigorous "burst" tests and are convenient and economical.

Aluminium foil laminated with paper or plastic is frequently used as a heat – sealed membrane hermetically closing the container under a plastic screw cap. The resulting aluminium membrane provides excellent barrier properties preventing moisture or gas transmission and tamper evident seal.

Laminates

Laminates are combination of different plies put together to get some desired properties. These form thin films using minimum material and are cost effective. These however, are not eco – friendly as they make recycling very difficult. Any laminate may consist of a number of piles selected from paper, cellulose, films foils coatings, etc. depending upon:

- Availability of the material.
- Technical requirements.
- Cost of base material.
- Cost of lamination process, and the yield.

These may include paper/ PVdC. paper.PVdC, paper/LDPE/ surlyn, regenerated cellulose/LDPE etc. Two laminations, which are widely used, include paper/foil/polythene and paper/foil/surlyn. In this lamination the easy printability, strength and brilliance of paper (outer ply) have been put together with the barrier properties of foil (center ply) and heat sealability of surlyn (inner ply). These are used for strip and sachet packaging.

The main pharmaceutical applications of laminations include, strip packaging, blister packaging and sachet packaging. Films, foils or laminates are flexible, tamper-evident and encourage the unit-dose packaging. These packs including the blister and strip packs are reasonably child-resistant packs. The demand for blister packs in particular is shooting up and there is a great potential of growth in this particular area.

Metals

Metal containers are strong, relatively unbreakable, opaque, and impervious to water vapour, gases odour and bacteria. They are resistant to high and low temperatures. Due to their chemical reactivity, metals require the application of coating and lacquers to prevent chemical reaction and corrosion. Special coatings and coating techniques are being developed for this purpose. Metal containers being used in the pharmaceutical industry include small-elongated collapsible tubes and shallow drums. Metal containers have dominated themselves in the field of aerosols. Though glass, plastic and plastic coated glass aerosols are finding their own specialized application, metal aerosols are likely to retain the bulk of the market as long as cost advantages are offered. Aluminium foil is another metal extensively used in the pharmaceutical industry. It has been dealt with under foils heading in this chapter.

Metals are also used for the formation of closures. Some of these closures are similar to those used on glass and plastic containers e.g. plastic and metal screw closures and friction closures such as plug or slip lids. Other metal closures include lever lids and permanent mechanically seamed-on closures.

Tin plate was once used for pastille tins, ointments tins and various build-up containers for powders, tablets, capsules, etc. These were followed by the use of aluminium. The use of metals is reducing, as they cannot compete with the cost of plastic and glass. Some metals currently in used in the pharmaceutical packaging include tin plate, tin-free steel, aluminium, alloy of aluminium and stainless steel. Aluminium and its alloys are used for aerosol containers, rigid and collapsible tubes, shallow drawn containers, blisters and sachet packaging. Tin plate containers are used for the formation of built-up containers and shallow drawn containers. Stainless steel is widely used in pharmaceuticals for mixing vessels and manufacturing equipment. It is also used for packaging of high-pressure aerosols. Tin is occasionally used for collapsible tubes. Metal containers still play a very important role in packaging, accounting for one-quarter of the total sales of all packaging materials.

Closures

A closure forms the most critical component of a container, as far as the stability and compatibility of a product are concerned. Though the primary function of the closure system is to contain the contents, safety and security may be necessary to prevent hazards resulting in leakage, seepage, pilferage, loss of quality, purity, etc. by some source of contamination, impurity, etc. Thereby the containers should not give rise to undesirable interactions between the contents and the outside environment. Additionally, closures must allow the easy and safe administration of drugs. Closures may be required to be pierced with a needle as in the case of multi-dose parenterals. In such cases closures are made from elastomeric materials (rubber). There are four main types of rubber, which are used in pharmaceuticals.

- Butyl rubbers (co-polymers of isobutylene and isoprene or butadiene).
- Nitrile rubbers (butadiene-acrylonitrile co- polymers).
- Chloroprene rubbers (Neoprene; polymers of 1:4 chloroprenes).
- Silicon rubbers.

Butyl rubbers are cheap, chemical resistant, with desired aging properties and low water vapour/ air permeability. It is not good for oils. Nitrile rubbers are resistant to oil, heat treatment, vapour absorption and permeability but leaching and bactericide absorption is high. Neoprene rubbers are resistant to heat and oil with low water absorption and permeability as compared to natural rubbers. Silicon rubbers are costly but most heat resistant (about 250^0c) with very low water absorption and permeability, strength and aging properties. When piercing is not being done, the containers are made from plastics such as polyethylene or polypropylene. An aluminium cap may secure the rubber closures.

Closures may be achieved by a number of basic means or a combination of these including pressure, temperature and adhesion. Pressure type closures make use of mechanical pressure of atmospheric pressure. Mechanical pressure is utilized in screw closures, plug seal, lever lid, etc. Whereas atmospheric pressure is utilized in vacuum-sealed tin. Use of temperature is made in case of closures e.g. welding, heat seals or electrical heat sealing. Adhesion is made use of in closures in the form of solvents, adhesives, cold seal material, etc.

The materials employed for the formation of closures include:

- Metals like aluminium, aluminium alloys, tinplate, tin free steel, stainless steel.
- Glass is used for formation of stoppers.
- Rubbers and plastics (may be thermosetting type or thermoplastic type).

Plastic may be modified by use of coatings/ lacquers. Thermosetting plastics include phenol formaldehyde (PF), bakelite and urea formaldehyde (UF), Thermoplastic-based containers may be made from PS, LDPE, PPDPE, PP, and polythene of medium or high density.

Liners are generally used in the inner side of the containers. These are made from aluminium foil, tin foil, polythene, expended polyethylene (EPE), saran (PVdC) film, PVdC coating, PET film, solid polythene, solid PVC and PTFE, etc. The closures used in pharmaceuticals include rubber closures, caps or overseals, and some special type of closures.

(a) ***Rubber closures:*** Rubber consists of several ingredients, one which is an elastomer. Rubber compounds used in pharmaceutical packaging contain only limited number of ingredients, which are difficult to extract. These closures do not pose a problem and can used in contact with a large number of drug preparations.

Rubber closures for containers for aqueous parenteral preparations have been classified into two types, according to B.P. Type I closures are those which meet the stringent requirements and which are to be preferred: Type II closures are those, which have mechanical properties suitable for special use (e.g. multiple piercing), cannot meet the requirements as severs as those for the first category because of the chemical composition. Fragmentation test and self-sealability of closures of parenteral preparation is shown in Table 2.1 and 2.2 respectively.

(b) ***Caps and overseals:*** caps or overseals are used to secure the rubber closure to the container in order to maintain the integrity of the seal under normal conditions of handling and storage. These caps are usually made of aluminium and may be equipped with a plastic top to facilitate opening. These form tamper-evident packs. Aluminium caps from secondary packaging material as these are not in direct contact with the product.

Table 2.1 Fragmentation test (IP 1996) for rubber closures

Place a volume of water corresponding to nominal volume minus 4ml in each of 12 clean vials

↓

Close vial with closure and secure caps for 16 hours

↓

Pierce the closures with 21 SWG hypodermic needle (bevel angle of 10^0 to 14^0) and inject 1ml water and remove 1ml air

↓

Repeat the above operation 4 times for each closure (use new needle for each closure)

↓

Count the number of fragments visible to the naked eye

↓

Total number of fragments should not be more then 10 except butyl rubber where the fragment should not exceed 15

Table 2.2 Self sealability (IP 1996) test for rubber closures applicable to multidose containers only

Fill 10 vials with water to nominal volume and close the vials with closure, secure the cap

↓

Pierce the cap 10 times at different sites with 21 SWG hypodermic needle.

↓

Immerse the vials in 0.1% w/v solution of methylene blue under reduced external pressure (27K Pa) for 10 minutes

↓

Restore the normal pressure and keep the container immersed for 30 minutes

↓

Wash the vials. None of the vials should contain trace of colored solution

(c) **Special types of closures:** Demographic trends pose new problems for packaging designers. While child-resistant closures safeguard children against drug intoxication, opening such packaging may prove to be difficult for the increasing number of elderly in the population

Tamper-Evident closures

Tampering includes three aspects, namely altering, pilfering and falsifying the pharmaceutical product. Tamper evident closures are designed to prevent accidents and malicious tampering to create safe packaging. The concept of tamper-evident packaging is increasingly being recognized in pharmacopoeias. These include products ranging from OTC drugs, toothpaste, topical and dermatological products, oral cosmetic liquids, contact lens solutions and tablets. In May 1992, the FDA listed 11 technologies capable of satisfying the definition of tamper-evident packaging, while a twelfth was added for sealed cartons. Various types of temper evident packaging are listed in Table 2.3.

Table 2.3 Tamper-evident packaging

• Film wrappers	• Breakable cap ring systems
• Blister packs	• Sealed tubes
• Bubble packs	• Plastic blind end sealed tubes
• Heat shrunk bands or wrappers	• Sealod cartons
• Paper foil for plastic packs	• Aerosol containers, and
• Bottles with inner mouth seals	• Metal and composite cans
• Tape seals	

Child Resistant closures

Tragic accidents involving the drug intoxication of children has led to new legislation making it difficult for drug packaging to be opened by young children, while being easier for the adults to open. These are termed as child resistant closures. These have proved effective in reducing child mortality from intoxication by oral prescription drugs. International standards have been laid down by ISO and ECN (European Committee for standardization for child resistant closures). The three most common reclosable child resistant closures are press-turn, squeeze-turn and a combination lock.

Closure efficiency

The ability of a closure to prevent the exchange between a product and the atmosphere outside is determined by keeping the pack with the closure under varying temperature and humidity conditions and evaluating the system These include:

- Placing a desiccant in a pack stored under high RH and evaluating the moisture gain.
- Putting liquid in the pack, placing it under high temperature and low relative humidity (RH) and then detecting the moisture loss.
- Holding pack under water with vacuum applied on the pack and observing the water ingress into the container.
- Checking a poor seal by applying vacuum on inverted pack and looking for seepage.
- Cap removal torque is checked on time-temperature related basis.
- Checking the compression ring seal in a cap liner.

II Secondary packaging materials

Paper and board packaging materials

Paper and board are broadly cellulose/ natural fiber materials. They find more diverse applications when compared with primary packaging materials. Cartons/corrugated board made from these materials is relatively recent invention.

Paper

Paper is widely used in pharmaceutical industry primarily due to the advantages like low cost, wide availability, non-toxic origin, availability in a variety of range, ease of printing, coating, lamination, opening, use, modification by addition of additives, good rigidity, strength, biodegradability and disposal.

It suffers from some drawbacks of being moisture sensitive, needs modification in permeability to moisture and gases, is porous and is unable to form a barrier. However, advantages of paper outweigh its limitations and it is used widely for formation of labels, leaflets, etc. Strength of paper depends on its GSM (gram per square meter). This serves as an important parameter for its quality control.

Paper and board may be used for applications such as wrapping materials. labels, patient package inserts (PPI), bags and sacks, corrugated boxes, cartons, shipment containers, gummed tape, fitment for cases, paper liners, lining and lamination. Without these, packaging of products would be more expensive and difficult.

Labels are slips made generally of paper and are attached to a container bearing name, description and other details regarding the product. Another feature being introduced into the pharmaceutical label is bar coding. EAN (European Article Numbering) bar coding has become an integral part of label for products competing in the global market. This coding is a facilitation tool to track the products effectively at any point of time to ensure effective supply chain management and speedy product recall in need There is an increasing trend of including information leaflets into products. Patient package inserts (PPI) are such leaflets, which are inserted into the product package. These contain non-technical explanation informing patients about the medicine in the package.

Paper and board containers include:

(a) Fiberboard kegs and drams, which are used for solid bulk drugs, bulk tablets or bulk excipient containment and transport.

(b) Paper and composite open mouth sacks used for bulk excipients where there is no risk to or from the environment.

(c) Composite containers that can be used as tubes for protection or for small powder drums.

Paper board and other cellulosic material have a major future in pharmaceutical industry. These are renewable and economical alternative to petroleum-based materials.

III Tertiary packaging Materials

Tertiary packaging Materials is used for bulk handling and shipping. It includes barrel, crate container, edge protector, Intermediate bulk container, pallets, slip-sheet, stretch wrap, flexible intermediate bulk container, big bag, "bulk bags", or "super sacks". Tertiary packaging is least concern with the quality of pharmaceuticals, it mainly helps in bulk transportation and so there is no need to discuss in details about it.

CHAPTER 3

Polymers and Plastics in Packaging

Summary

3.1 Introduction

Plastic can cover many synthetic or semi synthetic polymerization products. They are composed of organic condensation or addition polymers and may contain other substances to improve performance or economics. There are few natural polymers generally considered to be "plastics". Plastics can be formed into objects or films or fibers. Their name is derived from the fact that many are malleable, having the property of plasticity

3.2 Plastics and Polymers

Plastic can be classified in many ways, but most commonly by their polymer backbone are

- **Cellulose based Plastics:** In the nineteenth century a plastic material based on chemically modified natural polymers was discovered: it was first made by mixing pyroxylin, a partially nitrated form of cellulose (cellulose is the major component of plant cell walls), with alcohol and camphor.
- **Polyethylene:** Polyethylene is a polymer consisting of long chains of the monomer ethylene. Polyethylene is classified into several different categories based mostly on its density and branching: UHMWPE, HMWPE, HDPE, HDXLPE, PEX, MDPE, LDPE, LLDPE and VLDPE.
- **Polypropylene:** Polypropylene or polypropene (PP) is a thermoplastic polymer, made by the chemical industry and used in a wide variety of applications, including food packaging, ropes, textiles, plastic parts and reusable containers of various types.

- **Polystyrene:** Polystyrene is a polymer made from the monomer styrene, a liquid hydrocarbon that is commercially manufactured from petroleum by the chemical industry.
- **Polyethylene terephthalate (PTE):** It is a thermoplastic polymer resin of the polyester family and is used in synthetic fibers; beverage, food and other liquid containers; thermoforming applications; and engineering resins often in combination with glass fiber.
- **Polytetrafluoroethylene:** It is a synthetic fluoropolymer which finds numerous applications. PTFE's most well known trademark in the industry is the DuPont brand name Teflon, as DuPont initially discovered PTFE.
- **Polyurethane:** It is any polymer consisting of a chain of organic units joined by urethane links.
- **Polyvinylidene chloride:** It is a polymer derived from vinylidene chloride.
- **Polystyrene and PVC:** It is a new forms of plastics: "polystyrene" (PS) and "polyvinyl chloride" (PVC), developed by IG Farben of Germany.
- **Nylon:** The real star of the plastics industry in the 1930s was "polyamide" (PA), far better known by its trade name, "nylon". It was very strong but also very flexible. The first application was for bristles for toothbrushes.
- **Acrylonitrile butadiene Styrene(ABS):** It is a common thermoplastic used to make light, rigid, molded products.
- **Synthetic rubber**: A polymer that was critical to the war effort was "synthetic rubber", which was produced in a variety of forms. Synthetic rubbers are not plastics. Synthetic rubbers are elastic materials.

3.3 Drug-plastic consideration

Following are some categories of drug-plastic consideration:

Permeation	Leaching
Sorption	Chemical reaction

Alteration in the physical properties of plastic or product (Modification).

3.1 Introduction

Plastic can cover many synthetic or semi synthetic polymerization products. They are composed of organic condensation or addition polymers and may contain other substances to improve performance or economics. There are few natural polymers generally considered to be "plastics". Plastics can be formed into objects or films or fibers. Their name is derived from the fact that many are malleable, having the property of plasticity.

Some of very specific characteristics of plastics like ease with which they can be created, their high quality and freedom to design in various sizes and shapes, make it acceptable for vital applications in pharmaceutical packaging. Compared to glass, plastic containers are resistant to breakage thus offer safety. Pharmaceutical plastic containers primarily made of one or more polymers together with certain additives. Over all plastics is used as:

- Containers for transfusion solution and retention enemas.
- Containers for eye and Ears drops, Nasal sprays.
- Eyes ointment tubes.
- Spray bottles – squeeze bottle
- Suppositories pack
- Flexible packing
- Bottles, Jars, and Closures.

3.2 Plastics and Polymers

Plastic can be classified in many ways, but most commonly by their polymer backbone are :

- Polyvinyl chloride
- Polyethylene
- Polyurethanes
- Silicones
- Polymethyl methacrylate and other acrylics etc.
- Other classifications include
- Thermoplastic
- Thermoset
- Elastomer
- Engineering plastic
- Addition or Condensation (Polyaddition Plastics) etc.

Some plastics are partially crystalline and partially amorphous in molecular structure, giving them both a melting point (the temperature at which the attractive intermolecular forces are overcome) and one or more glass transitions (temperatures above which the extent of localized molecular is substantially increased). So-called **semi-crystalline plastics** includes:

- Polyethylene
- Polypropylene
- Poly vinyl chloride,
- Polyamides (nylons)
- Polyesters
- Some polyurethanes.

Many plastics are completely **amorphous**, such as Polystyrene and its copolymers, poly (methyl methacrylate), and all thermoset. A classification chart for plastic as packaging material is illustrated below:

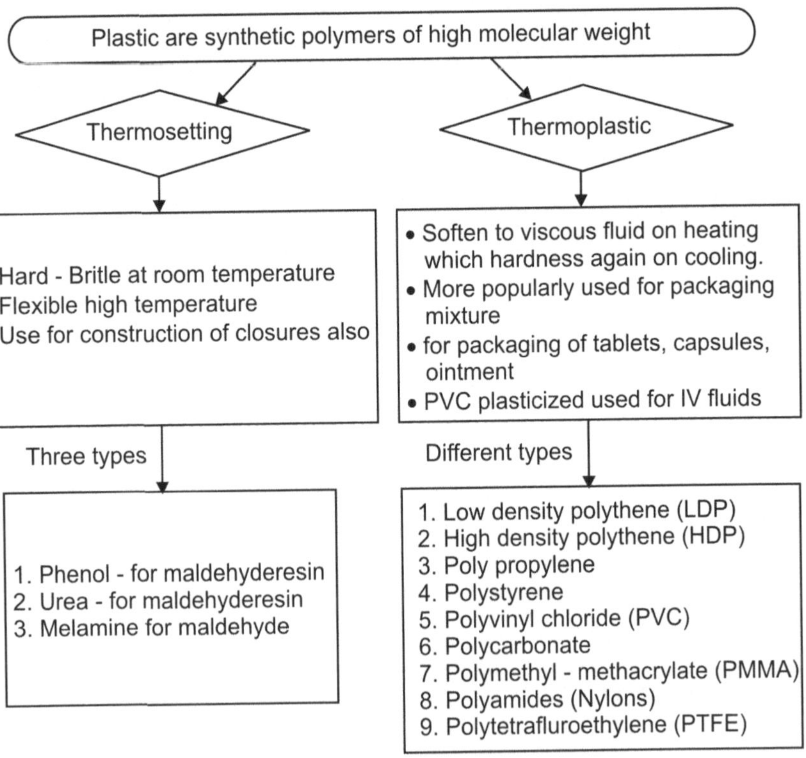

Plastics are polymers: long chains of atoms bonded to one another. Common thermoplastics range from 20,000 to 500,000 in molecular weight, while thermoset are assumed to have infinite molecular weight. These chains are made up of many repeating molecular units, known as "repeat units", derived from "monomers"; each polymer chain will have several 1000's of repeat units. The vast majority of plastics are composed of polymers of carbon and hydrogen alone or with oxygen, nitrogen, chlorine or sulfur in the backbone. (Some of commercial interest is silicon based.) The backbone is that part of the chain on the main "path" linking a large number of repeat units together. To vary the properties of plastics, both the repeat unit with different molecular groups "hanging" or "pendant" from the backbone, (usually they are "hung" as part of the monomers before linking monomers together to form the polymer chain). This customization by repeat unit's molecular structure has allowed plastics to become such an indispensable part of twenty first-century life by fine tuning the properties of the polymer.

To assist recycling of disposable items, the Plastic Bottle Institute of the Society of the Plastics Industry devised a now-familiar scheme to mark plastic bottles by plastic type. A recyclable plastic container using this scheme is marked with a triangle of three "chasing arrows", which enclose a number giving the plastic type:

Plastics type marks: the **Resin identification code**

1. **PET *(PETE)*:** Polyethylene Terephthalate-Commonly found on: 2-liter soft drink bottles, Bottles for Syrups, cooking oil bottles, peanut butter jars.
2. **HDPE**: High Density Polyethylene-Commonly found on: detergent bottles, milk jugs and Liquid oral preparations
3. **PVC**: Polyvinyl Chloride-Commonly found on: plastic pipes, outdoor furniture, shrink-wrap, water bottles, and salad dressing and liquid detergent containers.
4. **LDPE**: Low Density Polyethylene-Commonly found on: dry-cleaning bags produce bags, trash can liners, food storage containers, Shrink wraps, small bags for bulk packing of Tablets and capsules.

5. **PP**: Polypropylene-Commonly found on: bottle caps, drinking straws
6. **PS**: Polystyrene-Commonly found on: "Styrofoam peanuts," cups, plastic tableware, meat trays, take-away food clamshell containers.
7. **OTHER**: This plastic category, as its name of "other" implies, is any plastic other than the named 1-6, commonly found on: certain kinds of food containers, Tupperware, and Nalgene bottles.

People experimented with plastics based on natural polymers for centuries. In the nineteenth century a plastic material based on chemically modified natural polymers was discovered: Charles Goodyear discovered vulcanization of rubber (1839) and Alexander Parkes, English inventor (1813—1890) created the earliest form of plastic in 1855. He mixed pyroxylin, a partially nitrated form of cellulose (cellulose is the major component of plant cell walls), with alcohol and camphor. This produced a hard but flexible transparent material, which he called "**Parkesine**." The first plastic based on a synthetic polymer was made from phenol and formaldehyde, with the first viable and cheap synthesis methods invented by Leo Hendrik Baekeland in 1909, the product being known as **Bakelite.** Subsequently poly (vinyl chloride), polystyrene, polyethylene (polyethene), polypropylene (polypropylenc), polyamides (nylons), polyesters, acrylics, silicones, polyurethanes were amongst the many varieties of plastics developed and have great commercial success.

The development of plastics has come from the use of natural materials (e.g., chewing gum, shellac) to the use of chemically modified natural materials (e.g., natural rubber, nitrocellulose, collagen) and finally to completely synthetic molecules (e.g. Epoxy resins, polyvinyl chloride, polyethylene).

In 1959, Koppers Company in Pittsburgh, PA had a team that developed the expandable polystyrene foam cup. The foam cup can be buried, and it is totally stable such as is concrete and brick. No plastic film is required to protect the air and underground water. If it is burned with sufficient oxygen, the only chemicals generated are water and carbon dioxide. Burning a ton of cups results in less than 600 grams of ash. It can be recycled to make park benches, flowerpots and toys. Now we will discuss about individual plastic material in detail.

3.2.1 Cellulose-based plastics: celluloid and rayon

Celluloid/Xylonite proved extremely versatile in its field of application, providing a cheap and attractive replacement for ivory, tortoiseshell, and bone, and traditional products such as billiard balls and combs were much easier to fabricate with plastics. Hyatt was something of an industrial genius that understood what could be done with such a shapeable, or "plastic", material, and proceeded to design much of the basic industrial machinery needed to produce good-quality plastic materials in quantity. Some of Hyatt's first products were dental pieces, and sets of false teeth built around celluloid proved cheaper than existing rubber dentures. However, celluloid dentures tended to soften when hot, making tea drinking tricky, and the camphor taste tended to be difficult to suppress.

Bakelite (phenolic)

The limitations of celluloid led to the next major advance, known as "phenolic" or "phenol-formaldehyde" plastics. A chemist named Leo Hendrik Baekeland, a Belgian-born American living in New York State, was searching for insulating shellac to coat wires in electric motors and generators. Baekeland found that mixtures of phenol (C_6H_5OH) and formaldehyde (HCOH) formed a sticky mass when mixed together and heated, and the mass became extremely hard if allowed to cool and dry.

3.2.2 Polyethylene

Space-filling model of a **polyethylene** chain

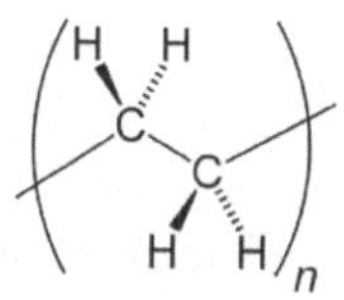

The repeating unit of **polyethylene**, showing its stereochemistry

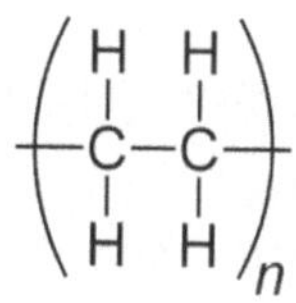

A simpler way of representing the repeating unit

Note, however, that the C–H bond angles are not 90° as this diagram would indicate, but are approximately 110°, since each carbon atom is tetrahedral (sp^3).

Polyethylene (IUPAC name polyethene) is a thermoplastic commodity heavily used in consumer products. Over 60 million tons of the materials are produced worldwide every year.

Description

Polyethylene is a polymer consisting of long chains of the monomer ethylene (IUPAC name ethene). The recommended scientific name 'polyethene' is systematically derived from the scientific name of the monomer. In certain circumstances it is useful to use a structure–based nomenclature. In such cases IUPAC recommends **poly(methylene)**. The difference is due to the 'opening up' of the monomer's double bond upon polymerisation.

In the polymer industry the name is sometimes shortened to **PE**. In the United Kingdom the polymer is commonly called **polythene**, although this is not recognized scientifically.

The ethene molecule (known almost universally by its common name ethylene), C_2H_4 is CH_2=CH_2, Two CH_2 groups connected by a double bond, thus:

Polyethylene is created through polymerization of ethene. It can be produced through radical polymerization, anionic addition polymerization, ion coordination polymerization or cationic addition polymerization. This is because ethene does not have any substituent groups that influence the stability of the propagation head of the polymer. Each of these methods results in a different type of polyethylene. Anti-oxidants and anti-static additives are used in bottle grades of polyethylene.

Classification of polyethylenes

Polyethylene is classified into several different categories based mostly on its density and branching. The mechanical properties of PE depend significantly on variables such as the extent and type of branching, the crystal structure, and the molecular weight.

- Ultra high molecular weight polyethylene (UHMWPE)
- High molecular weight polyethylene (HMWPE)
- High density polyethylene (HDPE)
- High density cross-linked polyethylene (HDXLPE)
- Cross-linked polyethylene (PEX)
- Medium density polyethylene (MDPE)
- Low density polyethylene (LDPE)
- Linear low density polyethylene (LLDPE)
- Very low density polyethylene (VLDPE)

Ultra high molecular weight polyethylene (UHMWPE)

It is polyethylene with a molecular weight numbering in the millions, usually between 3.1 and 5.67 million. The high molecular weight results in less efficient packing of the chains into the crystal structure as evidenced by densities less than high-density polyethylene (e.g. 0.935 - 0.930 g/cc).

The high molecular weight results in a very tough material. UHMWPE can be made through any catalyst technology, although Ziegler catalysts are most common. Because of its outstanding toughness, cut, wear and excellent chemical resistance, UHWMPE is used in a wide diversity of applications. These include can and bottle handling machine parts, moving parts on weaving machines, bearings, gears, artificial joints, edge protection on ice rinks, butchers' chopping boards.

High density polyethylene (HDPE)

It is defined by a density of greater or equal to 0.941 g/cc. HDPE has a low degree of branching and thus stronger intermolecular forces and tensile strength. HDPE can be produced by chromium/silica catalysts, Ziegler-Natta catalysts or metallocene catalysts. The lack of branching is ensured by an appropriate choice of catalyst (e.g. chromium catalysts or Ziegler-Natta catalysts) and reaction conditions. HDPE is used in products and packaging such as jugs, bottles, margarine tubs, garbage containers and water pipes.

HDPE is also widely used in the fireworks community. In tubes of varying length (depending on the size of the ordnance), HDPE is used as a replacement for the supplied cardboard mortar tubes for two primary reasons. One, it is much safer than the supplied cardboard tubes because if a shell were to malfunction and explode inside (*flower pot*) an HDPE tube, the tube will not shatter. The second reason is that they are reusable allowing designers to create multiple shot mortar racks. Pyrotechnicians discourage the use of PVC tubing in mortar tubes because it *will* shatter, sending shards of plastic at possible spectators, and will not show up in x-rays.

Cross-linked polyethylene (PEX)

It is a medium- to high-density polyethylene containing cross-link bonds introduced into the polymer structure, changing the thermoplast into an elastomer. The high-temperature properties of the polymer are improved, its flow is reduced and its chemical resistance is enhanced. PEX is used in some potable water plumbing systems, as tubes made of the material can be expanded to fit over a metal nipple, and it will slowly return to its original shape, forming a permanent, water-tight connection.

Medium density polyethylene (MDPE)

It is defined by a density range of 0.926 - 0.940 g/cc. MDPE can be produced by chromium/silica catalysts, Ziegler-Natta catalysts or metallocene catalysts. MDPE has good shock and drop resistance properties. It also is less notch sensitive than HDPE, stress-cracking resistance is better than HDPE. MDPE is typically used in gas pipes and fittings, sacks, shrink film, packaging film, carrier bags, screw closures.

Linear low density polyethylene (LLDPE)

It is defined by a density range of 0.915 – 0.925 g/cc. Is a substantially linear polymer, with significant numbers of short branches, commonly made by copolymerization of ethylene with short-chain alpha-olefins (e.g. 1-butene, 1-hexene, and 1-octene). LLDPE has higher tensile strength than LDPE. Exhibits higher impact and puncture resistance than LDPE. Lower thickness (gauge) films can be blown compared to LDPE, with better environmental stress cracking resistance compared to LDPE but is not as easy to process. LLDPE is used in packaging, particularly film for bags and sheets. Lower thickness (gauge) may be used compared to LDPE. Cable covering, toys, lids, buckets and containers, pipe. While other applications are available, LLDPE is used predominantly in film applications due to its toughness, flexibility, and relative transparency.

Low density polyethylene (LDPE)

It is defined by a density range of 0.910 – 0.940 g/cc. LDPE has a high degree of short and long chain branching, which means that the chains do not pack into the crystal structure as well. It has therefore less strong intermolecular forces as the instantaneous-dipole induced-dipole attraction is less. This results in a lower tensile strength and increased ductility. LDPE is created by free radical polymerization. The high degree of branches with long chains gives molten LDPE unique and desirable flow properties. LDPE is used for both rigid containers and plastic film applications such as plastic bags and film wrap.

Very low density polyethylene (VLDPE)

It is defined by a density range of 0.880 – 0.915 g/cc. Is a substantially linear polymer, with high levels of short chain branches, commonly made by copolymerization of ethylene with short-chain alpha-olefins (e.g. 1-butene, 1-hexene, and 1-octene). VLDPE is most commonly produced using metallocene catalysts due to the greater co-monomer incorporation exhibited by these catalysts. VLDPE's are used for hose and tubing, ice and frozen food bags, food packaging and stretch wrap, as well as impact modifiers when blended with other polymers.

Recently, much research activity has focused on the nature and distribution of long chain branches in polyethylene. In HDPE, a relatively small number of these branches, perhaps 1 in 100 or 1,000 branches per backbone carbon, can significantly affect the rheological properties of the polymer.

Ethylene copolymers: In addition to copolymerization with alpha-olefins, ethylene can also be copolymerized with a wide range of other monomers and ionic composition that creates ionized free radicals. Common examples include vinyl acetate (resulting product is ethylene-vinyl acetate copolymer, or EVA, widely used in athletic shoe sole foams), and a variety of acrylates (applications include packaging and sporting goods).

Physical properties

Depending on the crystallinity and molecular weight, a melting point and glass transition may or may not be observable. The temperature at which these occur varies strongly with the type of polyethylene. For common commercial grades of medium-density and high-density polyethylene, the melting point is typically in the range 120-130 °C. The melt point for average commercial low-density polyethylene is typically 105-115 °C. Most LDPE, MDPE, and HDPE grades have excellent chemical resistance and do not dissolve at room temperature because of the crystallinity. Polyethylene (other than cross-linked polyethylene) usually can be dissolved at elevated temperatures in aromatic hydrocarbons, such as toluene or xylene, or chlorinated solvents, such as trichloroethane or trichlorobenzene.

3.2.3 Polypropylene

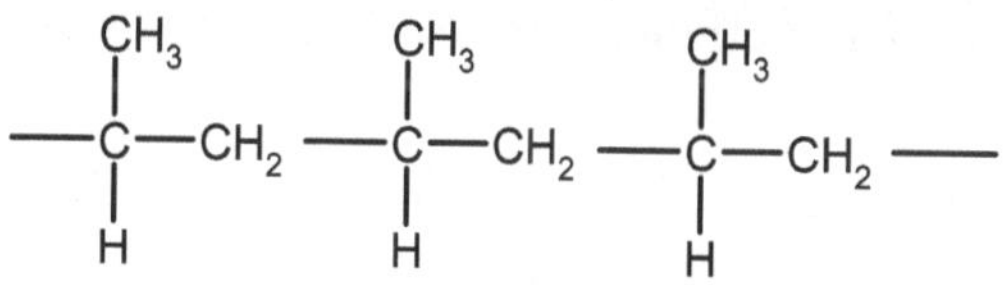

Chemical name : Poly(1-methylethylene)

Synonyms: Polypropylene; Polypropene; Polipropene 25 USAN]; Propene polymers; Propylene polymers; 1-Propene homopolymer

Chemical formula: $(C_3H_6)_x$

Monomer: Propylene (Propene)

Density: Amorphous: 0.85 g/cm^3

Crystalline: 0.95 g/cm^3

Melting point: ~ 165 °C

Glass transition temperature: -10 °C

Degradation point: 286 °C (559 K)

Resin identification Code: 

Polypropylene or polypropene (PP) is a thermoplastic polymer, made by the chemical industry and used in a wide variety of applications, including food packaging, ropes, textiles, plastic parts and reusable containers of various types, laboratory equipment, loudspeakers, automotive components, and polymer banknotes. Additionally it is used commonly in New Zealand and by the US Military as Thermal Underwear. An addition polymer made from the monomer propylene, it is rugged and unusually resistant to many chemical solvents, bases and acids.

3.2.4 Polystyrene

Density:	1050 kg/m^3
Specific Gravity	1.05
Electrical conductivity (σ)	10^{-16} S/m
Thermal conductivity	0.08 W/(m·K)
Young's modulus (E)	3000-3600 MPa
Tensile strength (σ_t)	46–60 MPa
Elongation at break	3–4%
Notch test	2–5 kJ/m^2
Glass temperature	95 °C
Melting point	240 °C
Heat transfer coefficient (λ)	0.17 W/(m·K)
Linear expansion coefficient (α)	8×10^{-5} /K
Specific heat (c)	1.3 kJ/(kg·K)
Water absorption (ASTM)	0.03–0.1
Decomposition	± 2000 years

Polystyrene is a polymer made from the monomer styrene, a liquid hydrocarbon that is commercially manufactured from petroleum by the chemical industry. At room temperature, polystyrene is normally a solid thermoplastic, but can be melted at higher temperature for molding or extrusion, then resolidified. Polystyrene is an aromatic polymer.

Polystyrene was accidentally discovered in 1839 by Eduard Simon, an apothecary in Berlin. From storax, the resin of *Liquidambar orientalis*, he distilled an oily substance, a monomer which he named styrol. Several days later Simon found that the styrol had thickened, presumably due to oxidation, into a jelly he dubbed styrol oxide ("Styroloxyd"). By 1845 English chemist John Blyth and German chemist August Wilhelm von Hofmann showed that the same transformation of styrol took place in the absence of oxygen. They called their substance metastyrol. Analysis later showed that it was chemically identical to Styroloxyd. In 1866 Marcelin

Berthelot correctly identified the formation of metastyrol from styrol as a polymerization process. About 80 years went by before it was realized that heating of styrol starts a chain reaction which produces macromolecules, following the thesis of German organic chemist Hermann Staudinger (1881 - 1965). This eventually led to the substance receiving its present name, polystyrene. The I.G. Farben company began manufacturing polystyrene in Ludwigshafen, Germany, about 1931, hoping it would be a suitable replacement for die cast zinc in many applications. Success was achieved when they developed a reactor vessel that extruded polystyrene through a heated tube and cutter, producing polystyrene in pellet form.

Pure solid polystyrene is a colorless, hard plastic with limited flexibility. It can be cast into molds with fine detail. Polystyrene can be transparent or can be made to take on various colors. It is economical and is used for producing plastic model assembly kits, plastic cutlery, cases, and many other objects where a fairly rigid, economical plastic of any of various colors is desired.

Solid foam

Expanded polystyrene packaging material

Polystyrene's most common use, however, is as expanded polystyrene (EPS). Expanded polystyrene is produced from a mixture of about 90-95% polystyrene and 5-10% gaseous blowing agent, most commonly pentane or carbon dioxide. The solid plastic is expanded into a foam through the use of heat, usually steam. Extruded polystyrene (XPS), which is different from expanded polystyrene (EPS), is commonly known by the trade name Styrofoam. The voids filled with trapped air give it low thermal conductivity. This makes it ideal as a construction material and it is therefore sometimes used in structural insulated panel building systems. It is also used as insulation in building structures, as molded packing material for cushioning fragile equipment inside boxes, as packing "peanuts", as non-weight-bearing architectural structures (such as pillars), and also in crafts and model building, particularly architectural models. Foamed between two sheets of paper, it makes a more-uniform substitute for corrugated cardboard, trade named *Fome-Cor*. A more unexpected use for the material is as a lightweight fill for embankments in the civil engineering industry.

Expanded polystyrene used to contain CFCs, but other, more environmentally safe blowing agents are now used. Because it is an aromatic hydrocarbon, it burns with an orange-yellow flame, giving off soot, as opposed to non-aromatic hydrocarbon polymers such as polyethylene, which burn with a light yellow flame (often with a blue tinge) and no soot.

Production methods include sheet stamping (PS) and injection molding (both PS and HIPS).

The chemical makeup of polystyrene is a long chain hydrocarbon with every other carbon connected to a Phenyl group (an aromatic ring similar to benzene).

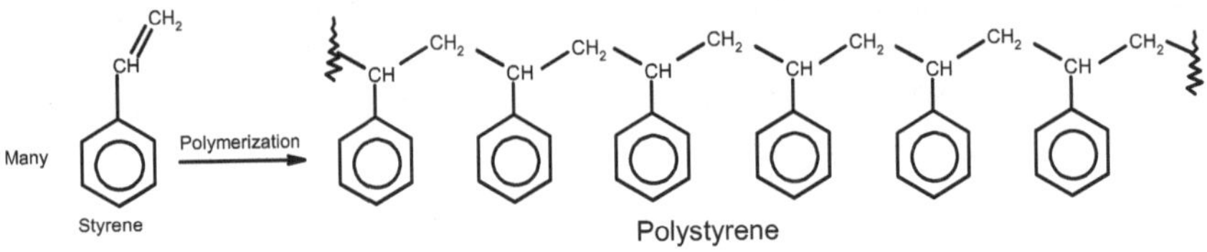

Metallocene-catalyzed polymerization can produce ordered *syndiotactic* polystyrene with the phenyl groups on alternating sides. This form is highly crystalline with a T_m of 270°C.

The resin identification code symbol for polystyrene is 

The majority of polystyrene products are currently not recycled due to a lack of suitable recycling facilities. Furthermore, when it is "recycled," it is not a closed loop — polystyrene cups and other packaging materials are usually recycled into fillers in other plastics, or other items that cannot themselves be recycled and are thrown away.

Pure polystyrene is brittle, but hard enough that a fairly high-performance product can be made by giving it some of the properties of a stretchier material, such as polybutadiene rubber. Depending upon the proportion of rubber added the shock resistance and toughness may be varied and accordingly these materials are often classified as intermediate-impact, high impact and super-impact polystyrenes. If polybutadiene is added during polymerization it can become chemically bonded to the polystyrene, resulting in **high-impact polystyrene** or **HIPS**. Common applications include use in toys and product casings. HIPS is usually injection molded in production. Autoclaving polystyrene can compress and harden the material.

Acrylonitrile butadiene styrene or ABS plastic is similar to HIPS: a copolymer of acrylonitrile and styrene, toughened with polybutadiene. Most electronics cases are made of this form of polystyrene, as are many sewer pipes. ABS pipes may become brittle over time.

Styrene can be copolymerized with other monomers; for example, divinylbenzene for cross-linking the polystyrene chains.

Expanded polystyrene is very easily cut with a hot-wire foam cutter, which is easily made by a heated and taut length of wire, usually nichrome due to nichrome's resistance to oxidation at high temperatures and its suitable electrical conductivity. The hot wire foam cutter works by heating the wire to the point where it can vaporize foam immediately adjacent to it. The foam gets vaporized before actually touching the heated wire, which yields exceptionally smooth cuts.

Polystyrene, **shaped and cut** with hot wire foam cutters, is used in architecture models, actual signage, amusement parks, movie sets, airplane construction, and much more. Such cutters may cost just a few dollars (for a completely manual cutter) to tens of thousands of dollars for large CNC machines that can be used in high-volume industrial production.

Polystyrene can also be cut with a traditional cutter. In order to do this without ruining the sides of the blade one must first dip the blade in water and cut with the blade at an angle of about 30°. The procedure has to be repeated multiple times for best results.

Polystyrene can also be cut on 3 and 5-axis routers, enabling large-scale prototyping and model-making. Special polystyrene cutters are available that look more like large cylindrical rasps.

This material is often sterilized post molding, either by irradiation or treatment with ethylene oxide. Post mold surface modification, usually with oxygen rich plasmas, is often done to introduce polar groups. Much of modern biomedical research relies on the use of such products; they therefore play a critical role in pharmaceutical research.

The health effects caused by consuming polystyrene when it migrates from food containers (primarily due to a leaching caused by heat exchange) into food is under serious investigation. Benzene, a material used in the production of polystyrene, is a known human carcinogen. Moreover, butadiene and styrene (in ABS), when combined, become benzene-like in both form and function.

Polystyrene is classified according to DIN4102 as a "B3" product, meaning highly flammable or "easily ignited". Consequently, though it is an efficient insulator at low temperatures, it is prohibited from being used in any exposed installations in building construction as long the material is not flame retarded e.g. with hexabromocyclododecane. It must be concealed behind drywall, sheet metal or concrete. Foamed plastic materials have been accidentally ignited and caused huge fires and losses. Examples include the Düsseldorf International Airport, the Channel tunnel, where it was inside a railcar and caught on fire, and the Browns Ferry Nuclear Power Plant, where fire reached through a fire retardant, reached the foamed plastic underneath, inside a firestop that did not consider bounding.

In addition to fire hazard, substances that contain acetone (such as most aerosol paint sprays), and cyanoacrylate glues can dissolve polystyrene.

Environmental concerns and bans on polystyrene

Expanded polystyrene is not easily recyclable due to its light weight and low scrap value. It is generally not accepted in curbside programs. Expanded polystyrene foam takes a very long time to decompose in the environment and has been documented to cause starvation in birds and other marine wildlife. According to the California Coastal Commission, it is a principal component of marine debris. A CIWMB (California Integrated Waste Management Board) Report finds that "in the categories of energy consumption, greenhouse gas effect, and total environmental effect, EPS's environmental impacts were second highest, behind aluminum."

The city of Berkeley, California was one of the first cities in the world to ban polystyrene food packaging (called Styrofoam in the media announcements). It was also banned in Portland, and Suffolk County, NY in 1990. Now, over 20 US cities have banned polystyrene food packaging, including Oakland, CA on Jan 1st 2007. San Francisco introduced a ban on the packaging on June 1st 2007.

A campaign to achieve the first ban of polystyrene foam from the food & beverage industry in Canada has been launched in Toronto as of January 2007, by local non-profit organization Naturo-Pack

3.2.5 Polyethylene Terephthalate (PET)

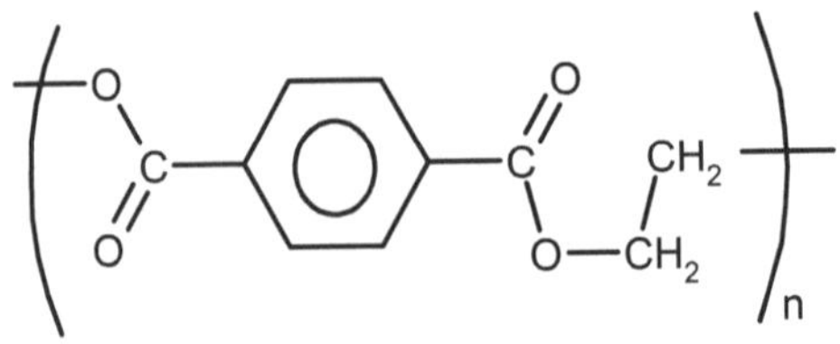

Chemical structure of polyethylene terephthalate

Density	1370 kg/m^3
Young modulus (E)	2800–3100 MPa
Tensile strength (σ_t)	55–75 MPa
Elongation @ break	50–150%
Glass temperature	75 °C
Notch test	3.6 kJ/m^2
Melting point	260 °C
Thermal conductivity	0.24 W/m.K
Linear expansion coefficient (α)	7×10^{-5}/K
Specific heat(c)	1.0 kJ/kg.K
Water absorption (ASTM)	0.16

Polyethylene terephthalate (PET, PETE) is a thermoplastic polymer resin of the polyester family and is used in synthetic fibers; beverage, food and other liquid containers; thermoforming applications; and engineering resins often in combination with glass fiber. It is one of the most important raw materials used in man-made fibers.

Depending on its processing and thermal history, it may exist both as an amorphous (transparent) and as a semi-crystalline (opaque and white) material. Its monomer can be synthesized by the esterification reaction between terephthalic acid and ethylene glycol with water as a byproduct, or the transesterification reaction between ethylene glycol and dimethyl terephthalate with methanol as a byproduct. Polymerization is through a polycondensation reaction of the monomers (done immediately after esterification/transesterification) with ethylene glycol as the byproduct (the ethylene glycol is recycled in production).

The majority of the world's PET production is for synthetic fibers (in excess of 60%) with bottle production accounting for around 30% of global demand. In discussing textile applications, PET is generally referred to as simply polyester while PET is used most often to refer to packaging applications.

It is manufactured under trade names **Arnite**, **Impet** and **Rynite**, **Ertalyte**, **Hostaphan Melinex** and **Mylar** films, and **Dacron**, **Diolen**, **Terylene** & **Trevira** fibers.

Uses of PET

PET can be semi-rigid to rigid, depending on its thickness, and is very lightweight. It makes a good gas and fair moisture barrier, as well as a good barrier to alcohol (requires additional "Barrier" treatment) and solvents. It is strong and impact-resistant. It is naturally colorless and transparent.

When produced as a thin film (often known by the trade name Mylar), PET is often coated with aluminium to reduce its permeability, and to make it reflective and opaque. PET bottles are excellent barrier materials and are widely used for soft drinks, (see carbonation). PET or Dacron is also used as a thermal insulation layer on the outside of the International Space Station. For certain specialty bottles, PET sandwiches an additional polyvinyl alcohol to further reduce its oxygen permeability.

When filled with glass particles or fibers, it becomes significantly stiffer and more durable. This glass-filled plastic, in a semi-crystalline formulation, is sold under the tradename **Rynite**, **Arnite**, **Hostadur** & **Crastin**.

PETE

Resin identification code

While all thermoplastics are technically recyclable, PET bottle recycling is more practical than many other plastic applications. The primary reason is that plastic carbonated soft drink bottles and water bottles are almost exclusively PET which makes them more easily identifiable in a recycle stream. PET has a resin identification code of 1. PET, as with many plastics, is also an excellent candidate for thermal recycling (incineration) as it is composed of carbon, hydrogen and oxygen with only trace amounts of catalyst elements (no sulfur) and has the energy content of soft coal.

PET was patented in 1941 by the Calico Printers' Association of Manchester. The PET bottle was patented in 1973.

One of the most important characteristics of PET is referred to as I.V. (intrinsic viscosity).

The I.V. of the material, measured in dl/g (deciliters/gram) is dependent upon the length of its polymer chains. The longer the chains, the stiffer the material, and therefore the higher the I.V. The average chain length of a particular batch of resin can be controlled during polymerization.

An I.V. of about:

0.60 - Would be appropriate for fiber

0.65 - Film

0.76-0.84 - Bottles

0.85 - Tire cord

PET is hygroscopic, meaning that it naturally absorbs water from its surroundings. However, when this 'damp' PET is then heated a chemical reaction known as hydrolysis takes place between the water and the PET which reduces its molecular weight (IV) and its physical properties. This means that before the resin can be processed in a molding machine, as much moisture as possible must be removed from the resin. This is achieved through the use of a desiccant or dryers before the PET is fed into the processing equipment.

Inside the dryer, hot dry air is pumped into the bottom of the hopper containing the resin so that it flows up through the pellets removing moisture on its way. The hot wet air leaves the top of the hopper and is first run through an after-cooler, because it is easier to remove moisture from cold air than hot air. The resulting cool wet air is then passed through a desiccant bed. Finally the cool dry air leaving the desiccant bed is re-heated in a process heater and sent back through the same processes in a closed loop. Typically residual moisture levels in the resin must be less than 40 ppm before processing. Dryer residence time should not be shorter than about four hours. This is because drying the material in less than 4 hours would require a temperature above 160 °C, at which level hydrolysis would begin inside the pellets before they could be dried out.

Copolymers of PET

In addition to pure (homopolymer) PET, PET modified by copolymerization is also available.

In some cases, the modified properties of copolymer are more desirable for a particular application. For example, cyclohexane dimethanol (CHDM) can be added to the polymer backbone in place of ethylene glycol. Since this building block is much larger (6 additional carbon atoms) than the ethylene glycol unit it replaces, it does not fit in with the neighboring chains the way an ethylene glycol unit would. This interferes with crystallization and lowers the polymer's melting temperature. Such PET is generally known as PETG (Polyethylene terephthalate glycol).

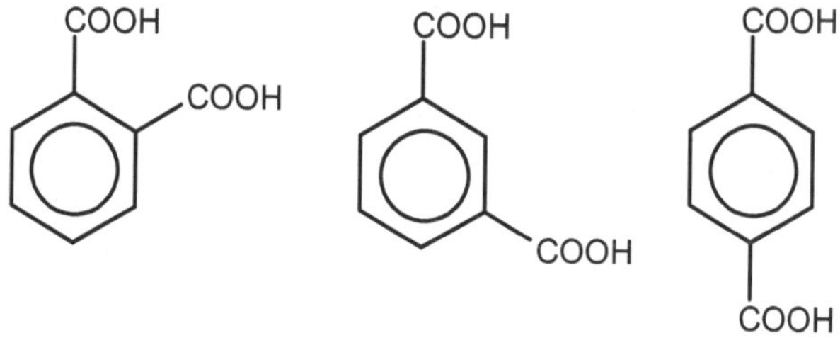

1, 2-phthalic acid 1,3-phthalic acid 1,4-phthalic acid

Replacing terephthalic acid (right) with isophthalic acid (center) creates a kink in the PET chain, interfering with crystallization and lowering the polymer's melting point.

Another common modifier is isophthalic acid, replacing some of the 1,4- (*para-*) linked terephthalate units. The 1,2- (*ortho-*) or 1,3- (*meta-*) linkage produces an angle in the chain, which also disturbs crystallinity.

Such copolymers are advantageous for certain molding applications, such as thermoforming, which is used to make tray or blister packages from PET sheet (sometimes called APET, for "amorphous PET"). On the other hand, crystallization is important in other applications where mechanical and dimensional stability are important, such as seat belts. For PET bottles, the use of small amounts of CHDM or other co-monomers can be useful: if only small amounts of co monomers are used, crystallization is slowed but not prevented entirely. As a result, bottles are obtainable via stretch blow molding ("SBM"), which are both clear and crystalline enough to be an adequate barrier to aromas and even gasses, such as the carbon dioxide in carbonated beverages.

Degradation of PET

PET is subject to various types of degradations during processing. The main degradations that can occur are hydrolytic, thermal and probably most important thermal oxidation. When PET degrades, several things happen: discoloration, chain scissions resulting in reduced molecular weight, formation of acetaldehyde and cross-links ("gel" or "fish-eye" formation). Discoloration is due to the formation of various cromophoric systems following prolonged thermal treatment at elevated temperatures. This becomes a problem when the optical requirements of the polymer are very high eg in packaging applications. Acetaldehyde is normally a colorless gas with a fruity smell. It forms naturally in fruit, but it can cause an off-taste in bottled water. Acetaldehyde forms in PET through the "abuse" of the material. High temperatures (PET decomposes above 300 °C or 572 °F), high pressures, extruder speeds (excessive shear flow raises temperature) and long barrel residence times all contribute to the production of acetaldehyde. When acetaldehyde is produced, some of it remains dissolved in the walls of a container and then diffuses into the product stored inside, altering the taste and aroma. This is not such a problem for non-consumables such as shampoo, for fruit juices, which already contain acetaldehyde or for strong-tasting drinks, such as soft drinks. For bottled water, low acetaldehyde content is quite important, because if nothing masks the aroma, even extremely low concentrations (10-20 ppb) of acetaldehyde can produce an off-taste. The thermal and thermo oxidative degradation results in poor procesibility characteristics and performance of the material.

One way to alleviate this is to use a copolymer. Comonomers such as CHDM or isophthalic acid lower the melting temperature and reduces the degree of crystallinity of PET (especially important when the material is used for bottle manufacturing). Thus the resin can be plastically formed at lower temperatures and/or with lower force. This helps to prevent degradation, reducing the acetaldehyde content of the finished product to an acceptable (that is, unnoticeable) level. See copolymers, above. Other ways to improve the stability of the polymer is by using stabilizers, mainly antioxidants such as phosphites. Recently, molecular level stabilization of the material using nanostructured chemicals has also been considered.

Antimony in PET

Antimony trioxide (Sb_2O_3) is a catalyst that is often used in the production of PET. It remains in the material and can thus in principle migrate out into food and drinks. Although antimony trioxide is of low toxicity, its presence is still of concern. The Swiss Federal Office of Public Health investigated the amount of antimony migration, comparing waters bottled in PET and glass: the antimony concentration of the water in PET bottles was higher, but still well below the allowed maximal concentrations. (report available in German and French only) The Swiss Federal Office of Public Health concluded that the health risk of these low concentrations is negligible (1% of the "tolerable daily intake" determined by the WHO).

Recrystallized PET

PET can be used to explore the crystallization of amorphous solids. The resin identification code can be used to verify the type of plastic is made of: many plastic beverage bottles have the letters PET or PETE and a code of 1 on the bottom, near the center. When a flame is held several inches below the bottle and slowly brought closer, part of the material will visibly change. This happens because high temperatures melt the PET. This releases the tension that was frozen in during the blow molding process and the polymer chains will shift to a more relaxed and disordered state, which results in shrinkage of the softened area. Because of the decreased order of the polymer chains, there are now fewer crystal nuclei. Consequently, when the crystallites re-form upon cooling they grow larger than the original crystallites in the bottle wall. Because the new crystallites are larger than the wave length of light, they will now cause light to scatter, giving the material an opaque white appearance.

Processing Equipment for PET containers

There are two basic molding methods, one-step and two-step. In two-step molding, two separate machines are used. The first machine injection molds the preform. The preform looks like a test tube. The bottle-cap threads are already molded into place, and the body of the tube is significantly thicker, as it will be inflated into its final shape in the second step using stretch-blow molding. In the second process, the preforms are heated rapidly and then inflated against a two-part mold to form them into the final shape of the bottle. Pre-forms (uninflated bottles) are now also used as containers for candy.

In one-step machines, the entire process from raw material to finished container is conducted within one machine, making it especially suitable for molding non-standard shapes (custom molding), including jars, flat oval, flask shapes etc. Its greatest merit is the reduction in space, product handling and energy, and far higher visual quality than can be achieved by the two-step system.

3.2.6 Polytetrafluoroethylene

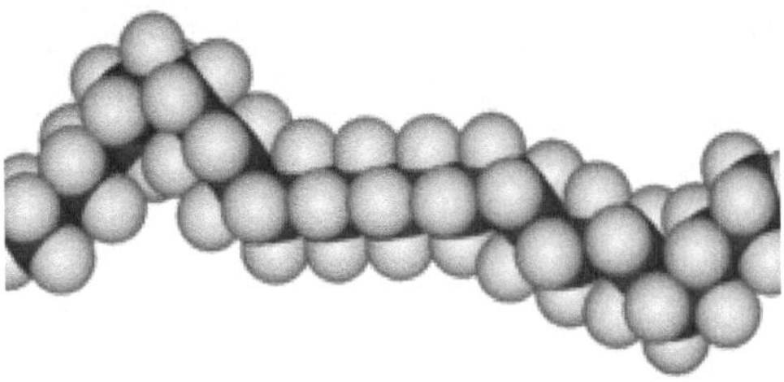

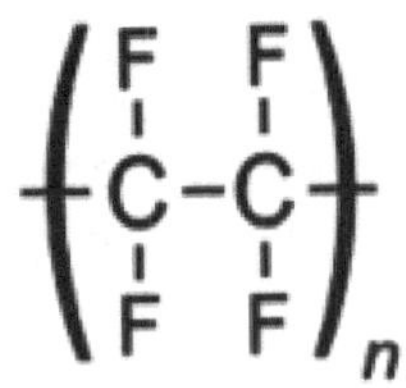

Systematic name	: poly(tetrafluoroethylene)
Other names:	:PTFE, (Teflon)
Empirical formula	: C_nF_{2n}
Density	: 2200 kg m^{-3}
Melting point	:327 °C
CAS number	:[9002-84-0]

Polytetrafluoroethylene (**PTFE**) is a synthetic fluoropolymer which finds numerous applications. PTFE's most well known trademark in the industry is the DuPont brand name **Teflon**, as DuPont initially discovered PTFE. PTFE has an extremely low coefficient of friction and is used as a non-stick coating for pans and other cookware. It is very non-reactive, and so is often used in containers and pipework for reactive and corrosive chemicals. Where used as a lubricant, PTFE significantly reduces friction, wear and energy consumption of machinery.

It was first sold commercially in 1946 and by 1950; DuPont was producing over a million pounds (450 t) per year in Parkersburg, West Virginia. In 1954, French engineer Marc Grégoire created the first pan coated with Teflon non-stick resin. An early advanced use was in the Manhattan Project as a material to coat valves and seals in the pipes holding highly reactive uranium hexafluoride in the vast uranium enrichment plant at Oak Ridge, Tennessee, when it was known as **K416**.

PTFE is a white solid at room temperature, with a density of about 2.2 g/cm^3. According to DuPont its melting point is 327 °C (620.6 °F), but its properties degrade above 260 °C (500 °F).

PTFE is often used to coat non-stick frying pans as it has very low friction and high heat resistance.

The coefficient of friction of plastics is usually measured against polished steel. Teflon's coefficient of friction is 0.1 or less, which is the lowest of any known solid material.

PTFE has excellent dielectric properties. This is especially true at high radio frequencies, making it suitable for use as an insulator in cables and connector assemblies and as a material for printed circuit boards used at microwave frequencies. Combined with its high melting temperature, this makes it the material of choice as a high-performance substitute for the weaker and lower melting point polyethylene that is commonly used in low-cost applications. Its extremely high bulk resistivity makes it an ideal material for fabricating long life electrets, useful devices that are the electrostatic analogues of magnets.

Due to its low friction, it is used for applications where sliding action of parts is needed: bearings, bushings, gears, slide plates, etc. In these applications it performs significantly better than nylon and acetal; it is comparable to **ultra high-molecular weight polyethylene** (UHMWPE), although UHMWPE is more resistant to wear than Teflon. For these applications, versions of teflon with mineral oil or molybdenum disulfide embedded as additional lubricants in its matrix are being manufactured.

Because of its chemical inertness, PTFE cannot be cross-linked like an elastomer. Powdered PTFE is used in pyrotechnic compositions as oxidizer together with powdered metals such as aluminum and magnesium (see Magnesium/Teflon/Viton). Upon ignition these mixtures form carbonaceous soot and the corresponding metal fluoride and release large amounts of heat. Hence they are used as infrared decoy flares and igniters for solid-fuel rocket propellants

PTFE is also used in body piercing, such as a sub-clavicle piercing, due to its flexibility and bio-compatibility.

Production of PTFE

PTFE is either synthesized by the emulsion polymerization of tetrafluoroethylene monomer under pressure, using free-radical catalysts,

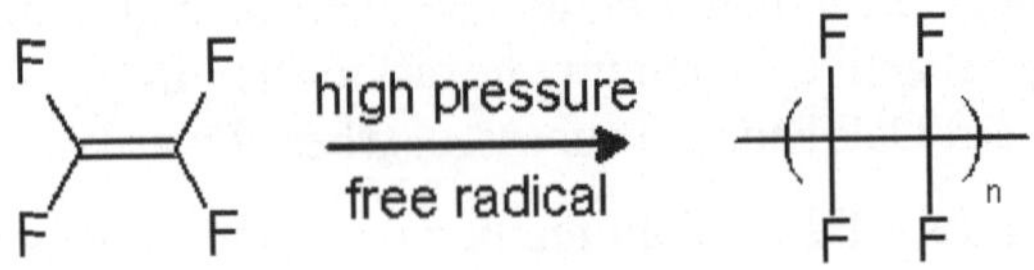

or it may be produced by the direct substitution of hydrogen atoms on polyethylene with fluorine, using polyethylene and fluorine gas at 20 °C :

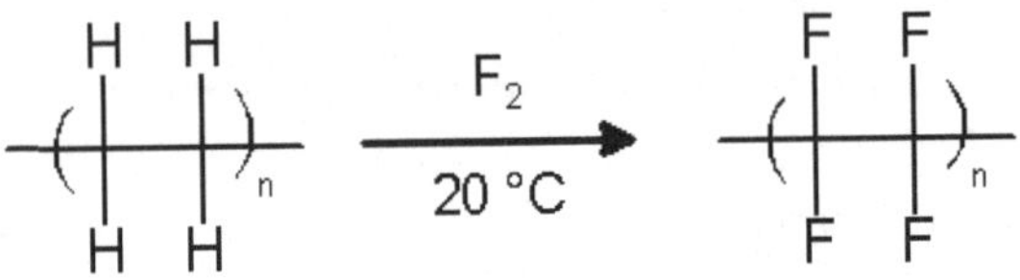

While PTFE itself is chemically inert and non-toxic, it begins to decompose above 660 °F (350 °C). These degradation products can be lethal to birds, and can cause flu-like symptoms in humans. A 1959 study, conducted before the Food and Drug Administration approved the material for use in food processing equipment, showed that the toxicity of fumes given off by the coated pan on dry heating was less than that of fumes given off by ordinary cooking oil.

In January 2006, DuPont, the only company that manufactures PFOA in the US, agreed to eliminate releases of the chemical from its manufacturing plants by 2015, but did not commit to completely phasing out its use of the chemical. This agreement is said to apply to not only PTFE used in cookware but also other products such as **food and pharmaceutical packaging**, clothing, and carpeting. PFOA is used only during the manufacture of the product—only a trace amount of PFOA remains after the curing process.

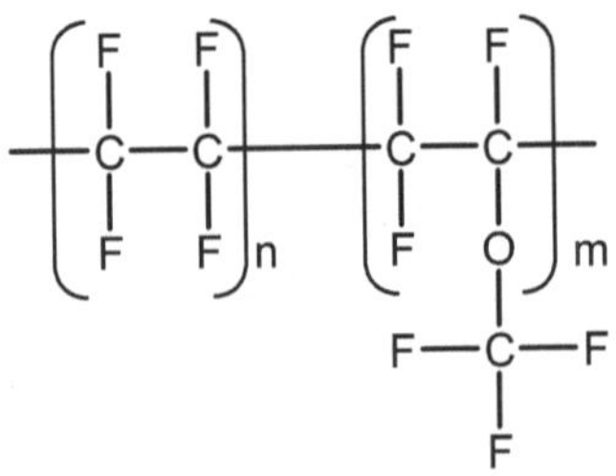

Teflon

Teflon is also used as the trade name for a polymer with similar properties, perfluoroalkoxy polymer resin (PFA).

Similar polymers

Other polymers with similar composition are known by the *Teflon* name: fluorinated ethylene-propylene (FEP) and perfluoroalkoxy polymer resin (PFA). They retain the useful properties of PTFE of low friction and non-reactivity, but are more easily formable. FEP is softer than PTFE and melts at 260 °C; it is highly transparent and resistant to sunlight.

3.2.7 Polyurethane

A **polyurethane** is any polymer consisting of a chain of organic units joined by urethane links. It is widely used in flexible and rigid foams, durable elastomers and high performance adhesives and sealants, fibers, seals, gaskets, condoms, carpet underlay, and hard plastic parts. Polyurethane products are often called "urethanes". They should not be confused with the specific substance urethane, also known as ethyl carbamate. Polyurethanes are not produced from ethyl carbamate, nor do they contain it.

A polyurethane polymer formed by reacting a diisocyanate with a polyol

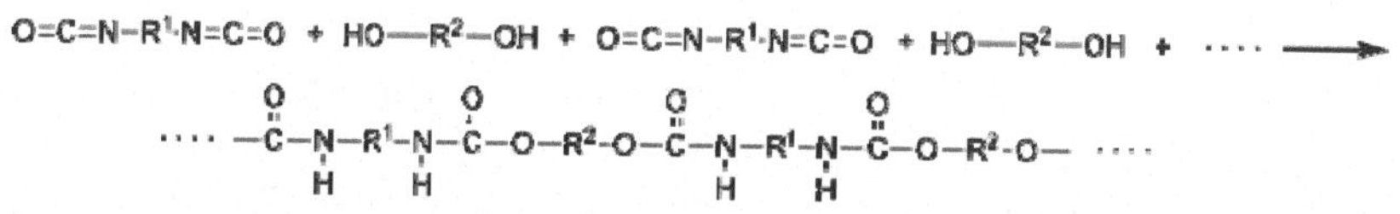

It was recognized that using the polyaddition principle to produce polyurethanes from liquid di-isocyanates and liquid polyether or polyester diols seemed to point to special opportunities, especially when compared to already existing plastics that were made by polymerizing olefins, or by polycondensation. Initially, work focused on the production of fibers and flexible foams. With development constrained by World War II, it was not until 1952 that polyisocyanates became commercially available. Commercial production of flexible polyurethane foam began in 1954, based on toluene diisocyanate (TDI) and polyester polyols. These materials were also used to produce rigid foams, gum rubber, and elastomers. Linear fibers were produced from hexamethylene diisocyanate (HDI) and 1,4-butanediol (BDO).

The first commercially available polyether polyol, poly(tetramethylene glycol), was introduced by DuPont in 1956. Less expensive polyalkylene glycols were introduced by BASF and Dow Chemical the following year, 1957. These polyether polyols offered technical and commercial advantages such as low cost, ease of handling, and better hydrolytic stability; and quickly supplanted polyester polyols in the manufacture of polyurethane goods. In 1960 more than 45,000 tons of flexible polyurethane foams were produced. As the decade progressed, the availability of chlorofluoroalkane blowing agents, inexpensive polyether polyols, and methylene diphenyl diisocyanate (MDI) heralded the development and use of polyurethane rigid foams as high performance insulation materials. In addition, automotive interior safety components such as instrument and door panels were produced by back-filling thermoplastic skins with semi-rigid foam.

Chemistry of Polyurethanes

generalized polyurethane reaction

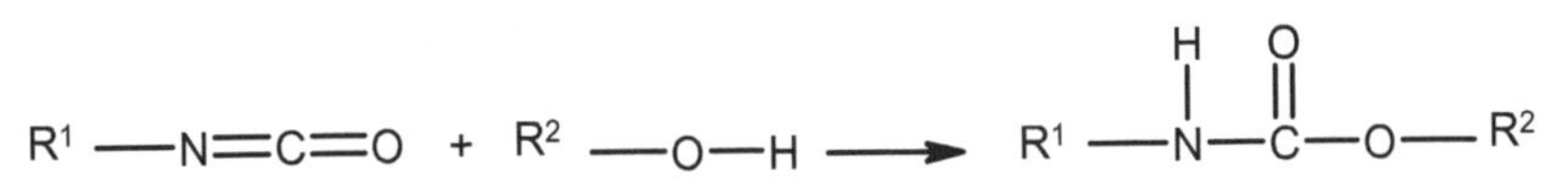

Polyurethanes are in the class of compounds called **reaction polymers,** which include epoxies, unsaturated polyesters, and phenolics. A urethane linkage is produced by reacting an isocyanate group, -N=C=O with a hydroxyl (alcohol) group, -OH. Polyurethanes are produced by the polyaddition reaction of a polyisocyanate with a polyalcohol (polyol) in the presence of a catalyst and other additives. In this case, a polyisocyanate is a molecule with two or more isocyanate functional groups, $R\text{-}(N{=}C{=}O)_{n \geq 2}$ and a polyol is a molecule with two or more hydroxyl functional groups, $R'\text{-}(OH)_{n \geq 2}$. The reaction product is a polymer containing the urethane linkage, -RNHCOOR'-. Isocyanates will react with any molecule that contains an active hydrogen. Importantly, isocyanates react with water to form a urea linkage and carbon dioxide gas; they also react with poly(ether)amines to form polyureas. Commercially, polyurethanes are produced by reacting a liquid isocyanate with a liquid blend of polyols, catalyst, and other additives. These two components are referred to as a polyurethane system, or simply a system. The isocyanate is commonly referred to in North America as the 'A-side' or just the 'iso'. The blend of polyols and other additives is commonly referred to as the 'B-side' or as the 'poly'. This mixture might also be called a 'resin' or 'resin blend'. In Europe the meanings for 'A-side' and 'B-side' are reversed. Resin blend additives may include chain extenders, cross linkers, surfactants, fire retardants, blowing agents, pigments, and fillers.

The first essential component of a polyurethane polymer is the isocyanate. Molecules that contain two isocyanate groups are called diisocyanates. These molecules are also referred to as monomers or monomer units, since they themselves are used to produce polymeric isocyanates that contain three or more isocyanate functional groups. Isocyanates can be classed as aromatic, such as diphenylmethane diisocyanate (MDI) or toluene diisocyanate (TDI); or aliphatic, such as hexamethylene diisocyanate (HDI) or isophorone diisocyanate (IPDI). An example of a polymeric isocyanate is polymeric diphenylmethane diisocyanate, which is a blend of molecules with two-, three-, and four- or more isocyanate groups, with an average functionality of 2.7. Isocyanates can be further modified by partially reacting them with a polyol to form a prepolymer. A quasi-prepolymer is formed when the stoichiometric ratio of isocyanate to hydroxyl groups is greater than 2:1. A true prepolymer is formed when the stoichiometric ratio is equal to 2:1. Important characteristics of isocyanates are their molecular backbone, % NCO content, functionality, and viscosity.

The second essential component of a polyurethane polymer is the polyol. Molecules that contain two hydroxyl groups are called diols, those with three hydroxyl groups are called triols, et cetera. In practice, polyols are distinguished from short chain or low-molecular weight glycol chain extenders and cross linkers such as ethylene glycol (EG), 1,4-butanediol (BDO), diethylene glycol (DEG), glycerine, and trimethylol propane (TMP). Polyols are polymers in their own right. They are formed by free radical addition of propylene oxide (PO), ethylene oxide (EO) onto a hydroxyl or amine containing initiator, or by polyesterification of a di-acid, such as adipic acid, with glycols, such as ethylene glycol or dipropylene glycol (DPG). Polyols extended with PO or EO are polyether polyols. Polyols formed by polyesterification are polyester polyols. The choice of initiator, extender, and molecular weight of the polyol greatly affect its physical state, and the physical properties of the polyurethane polymer. Important characteristics of polyols are their molecular backbone, initiator, molecular weight, % primary hydroxyl groups, functionality, and viscosity.

The polymerization reaction is catalyzed by tertiary amines, such as *dimethylcyclohexylamine*, and organometallic salts, such as *dibutyltindilaurate*. Furthermore, catalysts can be chosen based on whether they favor the urethane (gel) reaction, such as *diazobicyclooctane*, or the urea (blow) reaction, such as *bis-dimethylaminoethylether*, or specifically drive the isocyanate trimerization reaction, such as *potassium octoate*.

One of the most desirable attributes of polyurethanes is their ability to be turned into foam. Blowing agents such as water, and certain halocarbons, such as *HFC-245fa (1,1,1,3,3-pentafluoropropane)*, and hydrocarbons, such as *n-pentane*, can be incorporated into the poly side or added as an auxiliary stream. Water reacts with the isocyanate to create carbon dioxide gas, which fills and expands cells created during the mixing process. Halocarbons and hydrocarbons are chosen such that they have boiling points at or near room temperature. Since the polymerization reaction is exothermic, these blowing agents volatilize into a gas during the reaction process. They fill and expand the cellular polymer matrix, creating a foam. It is important to know that the blowing gas does not create the cells of a foam. Rather, they are formed during the mixing process as nucleating sites that the blowing gas fills and expands. In fact, high density microcellular foams can be formed without the addition of blowing agents by mechanically frothing or nucleating the poly blend prior to use.

Surfactants are used to modify the characteristics of the polymer during the foaming process. The are used to emulsify the liquid components, regulate cell size, and stabilize the cell structure to prevent collapse and surface defects. Rigid foam surfactants are designed to produce very fine cells and a very high closed cell content. Flexible foam surfactants are designed to stabilize the reaction mass while at the same time maximizing open cell content to prevent the foam from shrinking. The need for surfactant can be affected by choice of isocyanate, polyol, component compatibility, system reactivity, process conditions and equipment, tooling, shot weight, and part geometry.

Production of Polyurethanes

The main polyurethane producing reaction is between a diisocyanate (aromatic and aliphatic types are available) and a polyol, typically a polypropylene glycol or polyester polyol, in the presence of catalysts and materials for controlling the cell structure, (surfactants) in the case of foams. Polyurethane can be made in a variety of densities and hardnesses by varying the type of monomer(s) used and adding other substances to modify their characteristics, notably density, or enhance their performance. Other additives can be used to improve the fire performance, stability in difficult chemical environments and other properties of the polyurethane products.

Though the properties of the polyurethane are determined mainly by the choice of polyol, the diisocyanate exerts some influence, and must be suited to the application. The cure rate is influenced by the functional group reactivity and the number of functional isocyanate groups. The mechanical properties are influenced by the functionality and the molecular shape. The choice of diisocyanate also affects the stability of the polyurethane upon exposure to light. Polyurethanes made with aromatic diisocyanates yellow with exposure to light, whereas those made with aliphatic diisocyanates are stable. Softer, elastic, and more flexible polyurethanes result when linear difunctional polyethylene glycol segments, commonly called polyether polyols, are used to create the urethane links. This strategy is used to make spandex elastomeric fibers and soft rubber parts, as well as foam rubber. More rigid products result if polyfunctional polyols are used, as these create a three-dimensional cross-linked structure which, again, can be in the form of a low-density foam.

An even more rigid foam can be made with the use of specialty trimerization catalysts which create cyclic structures within the foam matrix, giving a harder, more thermally stable structure, designated as polyisocyanurate foams. Such properties are desired in rigid foam products used in the construction sector.

Polyurethane foam (including foam rubber) is usually made by adding small amounts of volatile materials, so-called blowing agents, to the reaction mixture. These can be simple volatile chemicals such as acetone or methylene chloride, or more sophisticated fluorocarbons which yield important performance characteristics, primarily thermal insulation.

Another common route to produce foams is the addition of water to one of the liquid precursors of polyurethane before they are mixed together. This reacts with a portion of the isocyanate, generating carbon dioxide throughout the liquid, creating relatively uniform bubbles which then harden to form a solid foam as polymerization progresses.

The presence of water means that a small proportion of reactions result in urea linkages —NC (=O) N—, rather than urethane linkages, so that the resulting material should technically be called poly (urethane-co-urea).

Careful control of viscoelastic properties — by modifying the catalysts and polyols used —can lead to memory foam, which is much softer at skin temperature than at room temperature.

There are then two main foam variants: one in which most of the foam bubbles (cells) remain closed, and the gas(es) remains trapped, the other being systems which have mostly open cells, resulting after a critical stage in the foam-making process (if cells did not form, or became open too soon, foam would not be created). This is a vitally important process: if the flexible foams have closed cells, their softness is severely compromised, they become pneumatic in feel, rather than soft; so, generally speaking, flexible foams are required to be open-celled.

The opposite is the case with most rigid foams. Here, retention of the cell gas is desired since this gas (especially the fluorocarbons referred to above) gives the foams their key characteristic: high thermal insulation performance.

A third foam variant, called microcellular foam, yields the tough elastomeric materials typically experienced in the coverings of car steering wheels and other interior automotive components.

Fully reacted polyurethane polymer is chemically inert. In the United States, no exposure limits have been established by OSHA (Occupational Safety and Health Administration) or ACGIH (American Conference of Governmental Industrial Hygienists). It is not regulated by OSHA for carcinogenicity. Polyurethane polymer is a combustible solid and will ignite if exposed to an open flame for a sufficient period of time. Decomposition products include carbon monoxide, oxides of nitrogen, and hydrogen cyanide. Polyurethane polymer dust can cause mechanical irritation to the eyes and lungs. Liquid resin blends and isocyanates may contain hazardous or regulated components. They should be handled in accordance with manufacturer recommendations found on product labels and in MSDS (Material Safety Data Sheet) and product technical literature. Isocyanates are known skin and respiratory sensitizers, and proper engineering controls should be in place to prevent exposure to isocyanate liquid and vapor.

In the United States, additional health and safety information can be found through organizations such as the Polyurethane Manufacturers Association (PMA) and the Center for the Polyurethanes Industry (CPI), as well as from polyurethane system and raw material manufacturers. Regulatory information can be found in the Code of Federal Regulations Title 21 (Food and Drugs) and Title 40 (Protection of the Environment).

Uses of Polyurethane

Polyurethane products have many uses. Over three quarters of the global consumption of polyurethane products is in the form of foams, with flexible and rigid types being roughly equal in market size. In both cases, the foam is usually behind other materials: flexible foams are behind upholstery fabrics in commercial and domestic furniture; rigid foams are inside the metal and plastic walls of most refrigerators and freezers, or behind paper, metals and other surface materials in the case of thermal insulation panels in the construction sector. Its use in garments is growing: for example, in lining the cups of brassieres. Polyurethane is also used for moldings which include door frames, columns, balusters, window headers, pediments, medallions and rosettes.

The precursors of expanding polyurethane foam are available in many forms, for use in insulation, sound deadening, flotation, industrial coatings, Pharmaceutical packing material, and even cast-in-place upholstery padding. Since they adhere to most surfaces and automatically fill voids, they have become quite popular in these applications. Polyurethane is used as a black wrapping for timepiece bracelets over the main material which is generally stainless steel. It is used for comfort, style, and durability.

3.2.8 Polyvinylidene chloride

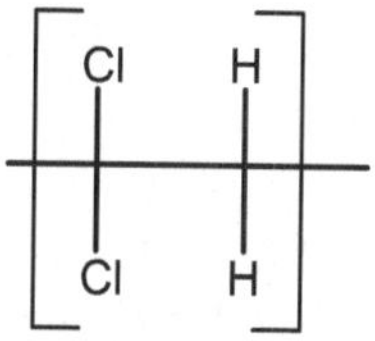

Polyvinylidene chloride

Polyvinylidene chloride is a polymer derived from vinylidene chloride. Ralph Wiley, a Dow Chemical lab worker, accidentally discovered polyvinylidene chloride in 1933. While cleaning laboratory glassware, Wiley came across a vial he could not scrub clean. Dow researchers made this material into a greasy, dark green film, first called "Eonite" and then Saran. The military sprayed Saran on fighter planes to guard against salty sea spray and carmakers used it for upholstery. Dow later got rid of the Saran's green color and unpleasant odor.

The most popularly known use of polyvinylidene chloride came in 1953, when Saran Wrap, a plastic food wrap was introduced. In 2004, however, the formula was changed to low density polyethylene out of environmental concerns about the chloride.

Saran fiber is manufactured by melt spinning vinylidene chloride copolymer. Saran is pigment dyed before fiberspinning if color is desired.

Saran Fiber is a heavy fiber. It is a **remarkable barrier** against water, oxygen and aromas. It has superior chemical resistance to alkalis and acids, it is insoluble in oil and organic solvents, has very low moisture regain and is impervious to mold, bacteria, and insects. Saran fiber has a high elastic recovery and resists wrinkling and creasing. Because Saran fiber is pigment dyed before fiberspinning, it has excellent colorfastness and high light permeability. Saran is also flame-retardant and self distinguishing, it may soften or char in flame, and decomposes in extreme heat.

Saran fiber comes in monofilament, multifilament-twist, and as a staple fiber. It is also available in thermochromic (color changing) and luminescent (glow in the dark) fibers.

Uses of polyvinylidene chloride

In packaging, polyvinylidene chloride is also applied as a water-based coating to other plastic films such as biaxially-oriented polypropylene (BOPP) and polyester (PET). This coating increases the barrier properties of the film, reducing the permeability of the film to oxygen and flavours and thus extending the shelf life of the food and medicaments inside the package. In household purposes it is used for Cleaning cloths, filters, screens, tape, shower curtains, garden furniture, food packaging. In industry it is used for Screens, artificial turf, waste-water treatment materials, underground materials. Other miscellaneous uses are like Doll hair, Barbie hair, stuffed animals, fabrics, fishnet, shoe insoles etc.

3.2.9 Polystyrene and PVC

After the First World War, improvements in chemical technology led to an explosion in new forms of plastics. Among the earliest examples in the wave of new plastics were "polystyrene" (PS) and "polyvinyl chloride" (PVC), developed by IG Farben of Germany.

Polystyrene is a rigid, brittle, inexpensive plastic that has been used to make plastic model kits and similar knickknacks. It would also be the basis for one of the most popular "foamed" plastics, under the name "styrene foam" or "Styrofoam". Foam plastics can be synthesized in an "open cell" form, in which the foam bubbles are interconnected, as in an absorbent sponge, and "closed cell", in which all the bubbles are distinct, like tiny balloons, as in gas-filled foam insulation and floatation devices. In the late 1950s "High Impact" styrene was introduced, which was not brittle. It finds much current use as the substance of toy figurines and novelties.

PVC has side chains incorporating chlorine atoms, which form strong bonds. PVC in its normal form is stiff, strong, heat and weather resistant, and is now used for making plum_bing, gutters, house siding, enclosures for computers and other electronics gear.

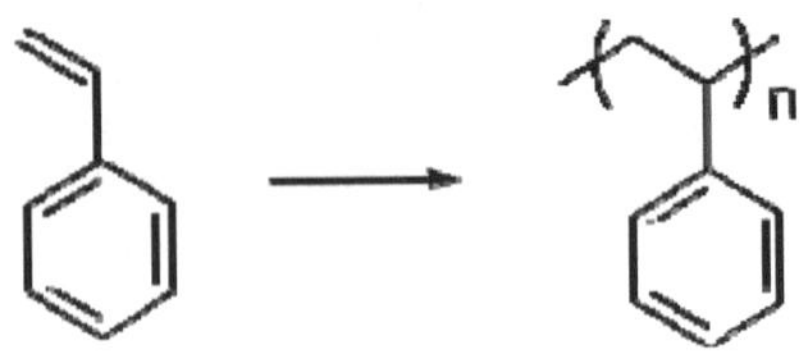

Polysryrene

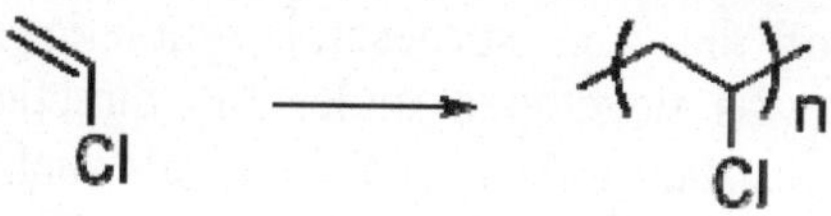

PVC

PVC can also be softened with chemical processing, and in this form it is now used for **shrink-wrap, food packaging**, and raingear.

3.2.10 Nylon

The real star of the plastics industry in the 1930s was "polyamide" (PA), far better known by its trade name, "nylon". It was very strong but also very flexible. The first application was for bristles for toothbrushes. However, Du Pont's real target was silk, particularly silk stockings. Carothers and his team synthesized a number of different polyamides including polyamide6.6 and 4.6., they also discovered polyesters.

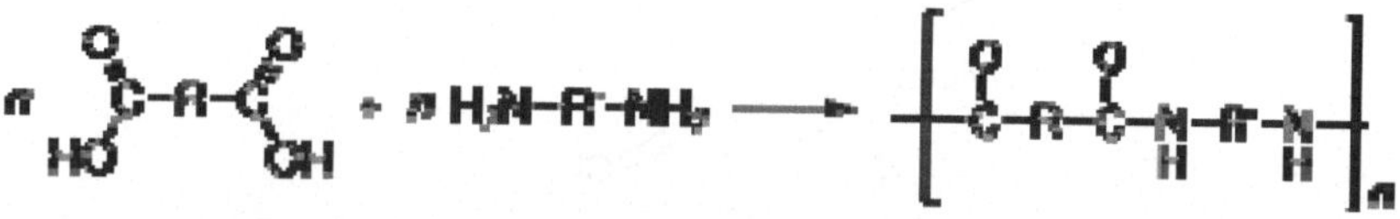

General condensation polymerization reaction for nylon

Nylon mania came to an abrupt stop at the end of 1941 when the USA entered World War II. The production capacity that had been built up to produce nylon stockings, or just "nylons", for American women was taken over to manufacture vast numbers of parachutes for fliers and paratroopers. Subsequently polyamides 6, 10, 11, 12 have been developed based on monomers which are ring compounds, e.g. caprolactam.

Nylons still remain important plastics, and not just for use in fabrics. In its bulk form it is very wear resistant, particularly if oil-impregnated, and so is used to build gears, bearings, bushings, and because of good heat-resistance, increasing for under-the-hood applications in cars, and other mechanical parts.

3.2.11 Synthetic rubber

A polymer that was critical to the war effort was "synthetic rubber", which was produced in a variety of forms. Synthetic rubbers are not plastics. Synthetic rubbers are elastic materials.

In 1931, one of the first successful synthetic rubbers, known as "neoprene", which was developed under the direction of E.K. Bolton. Neoprene is highly resistant to heat and chemicals such as oil and gasoline, and is used in fuel hoses and as an insulating material in machinery.

After the war, the Caltech researchers began to investigate the use of synthetic rubbers instead of asphalt as the fuel in the mixture. By the mid-1950s, large missiles were being built using solid fuels based on synthetic rubber, mixed with ammonium perchlorate and high proportions of aluminium powder. Such solid fuels could be cast into large, uniform blocks that had no cracks or other defects that would cause nonuniform burning. Ultimately, all large military rockets and missiles would use synthetic rubber based solid fuels, and they would also play a significant part in the civilian space effort.

3.2.12 Acrylonitrile butadiene styrene (ABS)

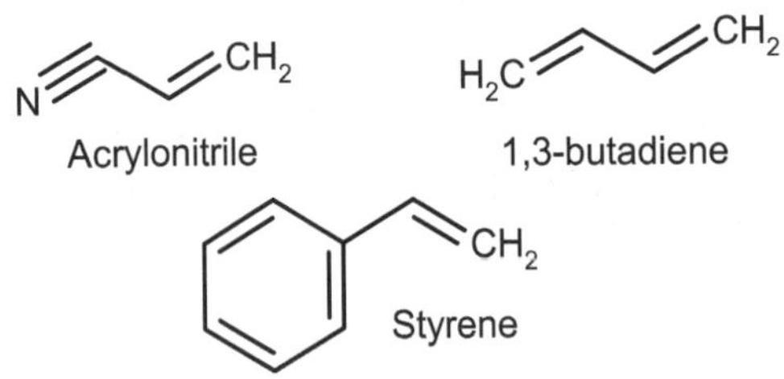

Monomers in ABS polymer

Acrylonitrile butadiene styrene, or **ABS**, (chemical formula $(C_8H_8 \cdot C_4H_6 \cdot C_3H_3N)_n$) is a common thermoplastic used to make light, rigid, molded products such as piping, musical instruments (most notably recorders), golf club heads (used for its good shock absorbance), automotive body parts, wheel covers, enclosures, protective head gear, vballs [reusable paintballs], and toys. In plumbing, ABS pipes are the black pipes (PVC pipes are white) and also in Plastic Pressure Pipe Systems. ABS plastic ground down to an average diameter of less than 1 micrometre is used as the colorant in some tattoo inks. Tattoo inks that use ABS are extremely vivid. This vividness is the most obvious indicator that the ink contains ABS, as tattoo inks rarely list their ingredients.

It is a copolymer made by polymerizing styrene and acrylonitrile in the presence of polybutadiene. The proportions can vary from 15 to 35% acrylonitrile, 5 to 30% butadiene and 40 to 60% styrene. The result is a long chain of polybutadiene criss-crossed with shorter chains of poly(styrene-co-acrylonitrile). The nitrile groups from neighbouring chains, being polar, attract each other and bind the chains together, making ABS stronger than pure polystyrene. The styrene gives the plastic a shiny, impervious surface. The butadiene, a rubbery substance, provides resilience even at low temperatures. ABS can be used between −25 and 60 °C.

Production of 1 kg of ABS requires the equivalent of about 2 kg of oil for raw materials and energy. It can also be recycled.

ABS is derived from acrylonitrile, butadiene, and styrene. Acrylonitrile is a synthetic monomer produced from propylene and ammonia; butadiene is a petroleum hydrocarbon obtained from butane; and styrene monomers, derived from coal, are commercially obtained from benzene and ethylene from coal. The advantage of ABS is that this material combines the strength and rigidity of the acrylonitrile and styrene polymers with the toughness of the polybutadiene rubber. The most amazing mechanical properties of ABS are resistance and toughness. A variety of modifications can be made to improve impact resistance, toughness, and heat resistance. The impact resistance can be amplified by increasing the proportions of polybutadiene in relation to styrene and acrylonitrile although this causes changes in other properties. Impact resistance does not fall off rapidly at lower temperatures. Stability under load is excellent with limited loads.

Even though ABS plastics are used largely for mechanical purposes, they also have good electrical properties that are fairly constant over a wide range of frequencies. These properties are little affected by temperature and atmospheric humidity in the acceptable operating range of temperatures. The final properties will be influenced to some extent by the conditions under which the material is processed to the final product; for example, molding at a high temperature improves the gloss and heat resistance of the product whereas the highest impact resistance and strength are obtained by molding at low temperature.

ABS polymers are resistant to aqueous acids, alkalis, concentrated hydrochloric and phosphoric acids, alcohols and animal, vegetable and mineral oils, but they are swollen by glacial acetic acid, carbon tetrachloride and aromatic hydrocarbons and are attacked by concentrated sulfuric and nitric acids. They are soluble in esters, ketones and ethylene dichloride.

The aging characteristics of the polymers are largely influenced by the polybutadiene content, and it is normal to include antioxidants in the composition. On the other hand, while the cost of producing ABS is roughly twice the cost of producing polystyrene, ABS is considered superior for its hardness, gloss, toughness, and electrical insulation properties. However, it will be degraded (dissolve) when exposed to acetone.

3.2.13 Aluminium oxynitride

Systematic name:	Aluminium oxynitride
Abbreviations:	AlON
Molecular formula:	$(AlN)_x.(Al_2O_3)_{1-x}$, mole fraction $0.30 \leq x \leq 0.37$
Appearance:	White or transparent solid
Density:	3.688 g/cm^3
Melting point:	~2140 °C
Boiling point:	(Al_2O_3~3000°C)
Solubility in water:	insoluble
Crystal structure:	cubic spinel

Aluminium oxynitride (AlON) is a transparent ceramic composed of aluminium, oxygen and nitrogen. It is marketed under the name ALON and described in U.S. Patent 4,520,116. The material remains solid up to 1200°C (2200°F), and is harder than glass. When formed and polished as a window, the material currently (2005) costs about US$10 to US$15 per square inch (~ US$20,000/m^2).

It is currently the crucial outer layer of experimental transparent armor being considered by the US Air Force for the windows of armored vehicles. Other applications include semiconductors and retail fixtures.

Objects are usually formed from pressed, cast or molded powder. The formed objects are then densified by heating in an oven, and polished until transparent. The method of polishing has a substantial positive effect on the armor's impact resistance.

3.3 Drug Plastic Considerations

In selecting a plastic container consideration must be given to:

- Composition of material.
- Its mechanical and chemical properties.
- Its ease of fabrication and printability.
- Requirement of rigidity or flexibility.
- Acceptance of plastic ingredients by F.D.C.

Depending on the type of plastic the total constituents may include the following:

- Basic polymer
- Residual monomers
- Plasticizers
- Accelerators
- Modifiers
- Stabilizers
- Fillers
- Colorants
- Antistatic agents
- Slip or Anti-slip additives
- U.V. absorbers.
- Lubricants
- Antioxidants

A container must protect the drug without altering the composition of the product start from the filling to the last removal of dose. Following are some categories of drug-plastic consideration:

1. Leaching
2. Permeation
3. Sorption
4. Chemical reaction
5. Alteration in the physical properties of plastic or product (Modification).

1. ***Leaching:*** The Leaching or migration of the plastic ingredients from the container to the drug creates a problem. Dyes may migrate into parenteral solution and cause a toxic effect.

2. ***Permeation:*** The permeations of gases, vapours or liquids through plastic packaging material can have an unfavorable effect on the shelf-life of a drug. If the dosage form is susceptible to hydrolysis and oxidation, it presents a problem. The permeation of vapours brings about the hydrolysis and oxidation of such products. The overall physical properties like aroma and taste of the product may be altered due to permeations of volatile matters through the walls of containers.

Formulations containing volatile ingredients might change their flavor because one or more of the ingredients are passing through the walls of the containers. In case if we pack emulsion in hydrophobic plastic container, the oils migrate and diffuse into plastic, which create problem in water in oil emulsion.

3. ***Sorption:*** The sorption reefers to the removal of the ingredients from the medicinal preparation by packaging material. The therapeutic efficacy of the preparation get decreased due to losses of potent medicament due to sorption.
4. ***Chemical reactivity:*** Plastic ingredients may react chemically with one or more components of medicinal product, which may alter the physical as well as chemical nature of the plastic or the drug product.
5. ***Modification:*** Any change in the physico-chemical properties of plastic containers by drug ingredients is referred to as modification. Permeation, sorption and leaching are the major factors, which lead to modification.

Comparative description of various polymers:

Types of Plastic	Properties	Application
1. Polyolefins Low Density Polyethylene (L.D.P.)	- Flexible very light, tough, - Impermeable to water vapours. - High permeability to gases - Permeable to oils preservative - Lack of transparency - Non – adherence of labels - Charged with static electricity.	Cosmetics, personal product, foods, tubes for topical ointment, bottles (Squeeze spray), Suppositories Pack
High Density Polyethylene (H.D.P.)	-More rigid, translucent, and inert -Easy handling and filling of containers -Highly Resistant to oils -Sterilized by autoclaving -Impermeable to water vapours	- Disposable syringes - Packaging of pharmaceutical preparations, infusion fluids, bottles, jars, closures, detergents, milk, food and cosmetics.
Polypropylene (P.P.)	Similar to H.D.P., but lighter, less opaque, more heat resistant, inert, impermeable to water vapours, flexible acquired original shape Low cost	- Disposable syringes - For dialysis fluids - Drug & Cosmetic - Syrup, juices, squeeze bottle - For retention enemas

Contd...

Types of Plastic	Properties	Application
(2) Polyvinylidene chloride (P.V.C.)	Less flexible than Polyethylene heaver More permeable to water vapours Less permeable to Gases than polyethylene High clarity	Bottles Jars Suppositories For retention enemas
(3) Polystyrene	Hard rigid Light material Easy to mould Permeable to water vapours Not sterilisable	Spoon Bottles Jars closure Nor suitable For sterile product
(4) Polytetrafluoroethylene (P.T.F.E.)	Excellent heart resistance Impermeable to water vapours Water vapours Permeable is nil Expensive Difficult to fabricate Resistance to solvents And chemicals	
(5) Polyamides (Nylon)	Very tough Non – transparent Permeability to water Vapours is high Heat resistance	Packaging films and laminates
(6) Polycarbonate	Transparent, sable, Heat resistant	

Evaluation of plastic containers

Biological test for plastic containers are perform by Extractive test as recommended by N.F. Physico-chemical test should be carried out on grounds of determine nonvolatile residue, residue on ignition, light transmission, permeation test. Containers should also be evaluated for chemical attack and sterilization subjects.

CHAPTER 4

Glass as a Packaging Material

Summary

4.1 Introduction

Glass has served the pharmaceutical industry for a couple of centuries, till the advent of plastics after which there has been a competition on the use of one with respect to the other. However, recently the advantages of both the containers have been put together made in union in the form of plastic coated glass. Advanced glass containers are light weight with a surface treatment. A number of glass containers used in pharmaceutical industry include ampoules, bottles, vials, syringes, and cartridges.

4.2 Types of Glasses

Several types of glass are available. These have been categorized as per USP into four different types. Type I glass is neutral glass and requires a higher working temperature and has a narrow working range thereby is difficult to process. Other glasses are type II, III and NP glass.

4.3 Glass properties

Glass properties can be calculated through statistical analysis of glass databases such as SciGlass and Interglad. If the desired glass property is not related to crystallization (e.g., liquids temperature) or phase separation, linear regression can be applied using common polynomial functions up to the third degree.

Chemical Properties

Sodium/Alkali Leaching, Acid Resistance (SR), Alkali Resistance (AR)

Electrical Properties

Volume Resistivity, Dielectric Constant, Surface Resistivity

Mechanical Properties

Stress, Young's Modulus, Poisson's Ratio, Modulus of Rupture, Shear Modulus, Bulk Modulus, Knoop Hardness, Density, Specific Gravity

Optical Properties

Refractive Index, Dispersion, Abbe Value, Transmission, Reflectivity, Birefringence Constant, Absorption

Thermal Properties

Coefficient of Thermal Expansion (CTE), Thermal Conductivity, Maximum Operating Temperature, Specific Heat, Surface Emissivity, Thermal Shock, Thermal Gradient, Working Temperature, Softening Point, Annealing Point, Strain Point, Normal Service, Extreme Limits, Thermal Shock, Temperature Gradient, Coefficient of Thermal Expansion, Tensile Strength, Design Tensile Strength

4.4 Glass as a liquid

Glass is generally treated as an amorphous solid rather than a liquid, though different views can be justified since characterizing glass as either 'solid' or 'liquid' is not an entirely straightforward matter. However, the notion that glass flows to an appreciable extent over extended periods of time is not supported by empirical research or theoretical analysis.

4.5 Colored Glasses

Amber is the fossilized resin of ancient pine trees, submerged under the sea in thin veins. It can be gathered along the North Sea coasts of East Anglia, south-west Jutland and the southern shores of the Baltic.

4.1 Introduction

Glass, discovered since 4,500 BC plays an indispensable part in our lives. Right from Cookery, light sources to high end technologies like computers and cameras, including Pharmaceutical packaging, the crucial component is glass. Glass is a uniform material of arguable phase (where the word "phase" is used to describe either a gas, liquid, or solid), usually produced when the viscous molten material cools very rapidly to below its glass transition temperature, without sufficient time for a regular crystal lattice to form. The most familiar form of glass is the silica-based material used for household objects such as light bulbs and windows. Glass is a biologically inactive material that can be formed into smooth and impervious surfaces. Under tension, glass is brittle and will break into sharp shards. Under compression, pure glass can withstand a great amount of force. The properties of glass can be modified or changed with the addition of other compounds or heat treatment.

Most glasses contain about 70–72 % by weight of silicon dioxide (SiO_2). The most common form of glass is soda-lime glass, which contains nearly 30 % sodium and calcium oxides or carbonates. Pyrex is borosilicate glass containing about 10 % boric oxide. Lead crystal is a form of lead glass that contains a minimum of 24 % lead oxide.

The major raw material of glass is sand (or "quartz sand") that contains almost 100 % of crystalline silica in the form of quartz. Although it is almost pure quartz, it may still contain a small amount (less than 1 %) of iron oxides that would color the glass, so this sand is usually depleted before production to reduce the iron oxide amount to less than 0.05 %. Large natural single crystals of quartz are pure silicon dioxide, and upon crushing are used for high quality specialty glasses. Synthetic amorphous silica, an almost 100 % pure form of quartz, is the raw material for the most expensive specialty glasses. The most common method for glass production is using molten tin, where the molten glass floats on top of the tin, thus giving it the name "float glass".

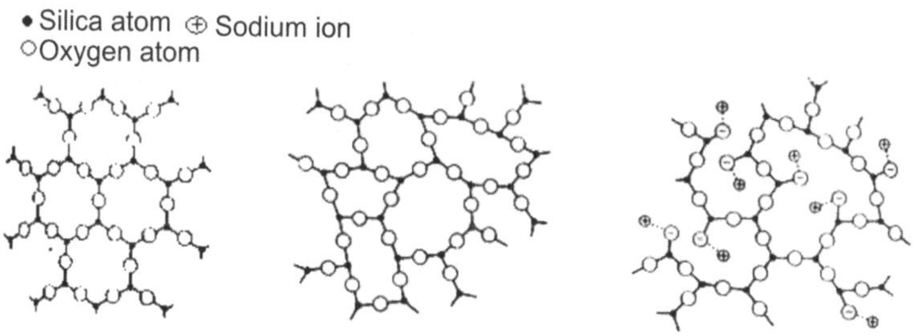

(a) Crystalline silica Ordered structure

(b) Silica glass Disordered structure

(c) Soda glass Disordered structure

Glass has served the pharmaceutical industry for a couple of centuries, till the advent of plastics after which there has been a competition on the use of one with respect to the other. However, recently the advantages of both the containers have been put together made in union in the form of plastic coated glass. Advanced glass containers are light weight with a surface treatment. A number of glass containers used in pharmaceutical industry include ampoules, bottles, vials, syringes, and cartridges.

Glass offers superior quality protection. It is economical, chemically inert impermeable with no diffusion or leakage (for good quality glass), strong and rigid, recyclable, hygienic, sterilizable, easily washable, resistant to high temperature, compatible and transparent. Another particular advantage that glass can offer is light protection.

Colorless white flint soda glass is made of silica, calcium oxide, sodium oxide, alumina and small quantities of ferric oxide, titanium dioxide, potassium and magnesium oxide.

Important dates in the History of Glass

~4500 B.C. to 3000 B.C. Glass first used in Egypt and Mesopotamia

~1500 B.C. First glass vessels made in Egypt

~200 B.C. Glass blowing invented in the near East

~600 A.D. Stained glass windows introduced in Constantinople

1674 A.D. Lead crystal invented by George Ravenscroft in England

1688 A.D. Cast plate glass-France

1690 A.D. First glassworks in Ireland

1827 A.D. Glass pressed into moulds-U.S.A.

1851 A.D. Crystal Palace built in London-first prefabricated building of iron and glass

1874 A.D. Glass toughened by oil quenching -France

1886 A.D. First semi-automated bottle maker -England

1903 A.D. First automatic bottle making machine invented by Michael Owens, an Irishman, in the U.S.A.

1903 A.D. Chance discovery of laminated glass by Benedictus in France

1915 A.D. Invention of borosilicate glass in the U.S.A.

1937 A.D. Photosensitive glass invented by Dalton in the U.S.A.

1957 A.D. Glass ceramic made by accident by Stookey in the U.S.A.

1959 A.D. Float glass process for plate glass introduced by Pilkingtons in the U.K.

4.2 Types of Glasses

Several types of glass are available. These have been categorized as per USP into four different types. Type I glass is neutral glass and requires a higher working temperature and has a narrow working range thereby is difficult to process. Other glasses are type II, III and NP glass (Table 4.1).

Table 4.1 Types of glass (As per USP).

General Description	Properties	Suitability
Type I Highly resistant, Borosilicate glass	Alkalinity is removed by using boric oxide to neutralize the oxides of potassium and Sodium	Preparations for parenteral administration
Type II Treated soda- lime Glass	Obtained by treating the hot surface of type III glass by sulphur dioxide/ammonium Sulphate/ ammonium chloride	Preparations for parenteral administration
Type III Soda-lime glass	It is an alkaline glass having high Percentage of lime and soda and No boric oxide as compare to Type I glass.	Not generally used for parenteral preparations until and unless indicated.
Type NP General purpose Soda-lime glass	It has similar compositions to that of Type III glass but there is no guarantee of similar properties.	Non parenteral articles e.g. oral or Topical use

Type I Glass (Borosilicate glass)

Borosilicate glass is a type of heat-resistant glass. Borosilicate glass was first developed by German glassmaker Otto Schott in the late 19th century and sold under the brand name "Duran" in 1893. After Corning Glass Works introduced Pyrex in 1915, it became a synonym for borosilicate glass in the English-speaking world. Holophane manufactures original equipment lenses for street lights under the Endural brand name. Bomex is a brand of borosilicate glassware made in China for a number of resellers in the United States and abroad. These have a high resistance to chemicals. In addition to the quartz, sodium carbonate, and calcium carbonate traditionally used in glassmaking, boron is used in the manufacture of borosilicate glass (Table 4.2). Typically, the resulting glass composition is about 70% silica, 10% boric oxide, 8% sodium oxide, 8% potassium oxide, and 1% calcium oxide. Though somewhat more difficult to make than traditional glass (Corning conducted a major revamp of their operations to make it), it is economical to produce because its superior durability, chemical and heat resistance finds excellent use in chemical laboratory equipment, cookware, lighting, and in certain cases, windows. Borosilicate glass is created by adding boron to the traditional glassmaker's "frit" of silicate sand, soda, and ground lime. Since Borosilicate glass melts at a higher temperature than ordinary silicate glass, some new techniques were required to bring it into industrial production. Borrowing from the welding trade, new burners combining oxygen with natural gas were required.

Borosilicate glass has a very low thermal expansion coefficient, about one-third that of ordinary glass. This reduces material stresses caused by temperature gradients, thus making it more resistant to breaking. This makes it a popular material for objects like telescope mirrors, where it is essential to have very little deviation in shape. It is also used in the processing of high-level nuclear waste, where the waste is immobilised in the glass through a process known as vitrification (contrast with Synroc).

Borosilicate glass begins to soften around 821 °C (1510 °F); at this temperature, the viscosity of type 7740 Pyrex is $10^{7.6}$ poise. Borosilicate glass is less dense than ordinary glass.

While more resistant to thermal shock than other types of glass, borosilicate glass can still crack or shatter when subject to rapid or uneven temperature variations. When broken, borosilicate glass tends to crack into large pieces rather than shattering (it will snap rather than splinter).

Optically, borosilicate glasses are crown glasses with low dispersion (Abbe numbers around 65) and relatively low refractive indices (1.51 - 1.54 across the visible range).

Table 4.2 Physical characters and Composition of Type I glass.

Density 2.23 g/cm^3

Mean Excitation Energy 134.0 eV

Element	Atomic number	Fraction
B	5	0.040064
O	8	0.539562
Na	11	0.028191
Al	13	0.011644
Si	14	0.377220
K	19	0.003321

Type II- Treated Soda-Lime glass

Type II is a soda glass, which has a surface treatment by sulphuring or sulphating, forming a coating of sodium sulfate, which neutralized the excessive surface alkalinity. Glasses are being given different treatments in order to improve their surface lubricity, increasing impact resistance and aesthetic appearance. Surface treated glass find wide application in pharmaceutical industry. **Quartz/Fused Silica,** this material is available from many sources and includes GE 124, and Corning 7980. There are many striate and bubble classes available. This is a material containing 96% silica that is used as telescope mirror substrates or other applications requiring good thermal and mechanical stability. It has a very low Coefficient of Thermal Expansion.

Type III- Soda Lime glass

Type III glass is the regular soda lime glass in which the containers are untreated and have an average chemical resistance. They have a limited alkalinityand generally not recommended for parenteral preparations until and unless indicated.

Type NP glasses (General-purpose soda lime glass)

Type NP glasses are the general-purpose soda lime glass and are supplied for non- parenteral product, intended for oral or topical use. Soda Lime is the standard type of glass available. All soda-lime glasses are defined as containing both sodium and calcium in their chemical make-up and are therefore subject to sodium leaching. All types of soda-lime glass can be heat strengthened, heat treated or chemically strengthened with some exceptions due to thickness constraints or coatings applied. Available from Abrisa in many forms:

- Clear (has a slight green tint)
- Clearview
- Superwite B-270
- Patterned or textured
- Tinted
- Non-glare etched or with various coatings
- Clear Soda-lime Glass

Clear, this is the same glass used in windows. It is produced by a process where the glass drawn over a molten tin bath to maintain very flat surfaces. It is available in many thicknesses from 0.5mm through 19mm and in many grades. Standard grade is available from 2mm through 19mm. Special grades are available from 0.5 through 3mm. These have tighter thickness control and are more expensive.

Clearview, this glass is has been specially formulated to eliminate the iron-causing green tint and therefore is much more transparent and pleasing to the eye. It is used where color rendition is critical. It is available in thicknesses from 3 to 12mm.

Patterned or textured, this glass is produced by drawing the glass between embossed rollers. It is available in a variety of standard patterns, but with limited thicknesses.

Tinted glass is standard clear glass to which special materials have been added during formation to produce a tinted glass. Colors include Green, Bronze, Blue, and various levels of Gray. They are limited in thicknesses and hue saturation (measured by transmission levels).

The British Pharmacopoeia categorizes the glass containers according to hydrolytic resistance into 4 types. Type I, neutral glass for parenteral use for human blood and blood components. Type II, high hydrolytic resistance glass suitable for acidic and neutral aqueous preparations for parenteral use with a pH lower than. Type III is soda lime glass with moderate hydrolytic resistance suitable for non- aqueous preparations for parenteral use, powder for parenteral use and for non-parenteral preparations. Type IV glass containers are soda lime silica glass with low hydrolytic resistance suitable for solid preparations that are not for parenteral use and for some liquids or semi-solid preparations that are not for parenteral use.

Glass is used for different containers ranging from ampoules, vials, cartridge tubes, disposable syringes and aerosol (Table 4.3). Ampoules are generally made up of neutral glass. These were one of the first unit dose containers but are increasingly being replaced by cartridge tubes and pre-filled syringes. Vials on the other hand are produced by soda glass, when used for tablets and capsules. Injection vials can still be obtained in either neutral or soda glass of occasionally in treated soda glass.

Table 4.3 Comparison of Glasses and their compositions.

Type	Composition	Remark	Application
Type – I OR Borosilicate Or Neutral glass OR Resistance	SiO– 80% Al_2O_3 – 95% Na_2O – 7% K_2 – O – 0.5% B_2O_3 – 12% CaO – 1%	Heat resistant chemically inert than soda – lime glass. High hydrolytic resistance leaching of NaOH only 0.5 PPM per year. Difficulty to melt and mould Very costing	Chemical glassware Ovenware Container for Alkali sensitive preparation
Type – II Treated sodalime glass (Soda – glass with surface treatment)-sulphured containers	SiO– 75% Al_2O_3 – 2% Na_2O – 10% K_2 – O – 0.5% B_2O_3 – 3% CaO – 10%	High hydrolytic resistance – due – to surface treatment (dealkalized by SO_2) Softer than borosilicate easy to mould Cheaper than Borosilicate	Containers for Alkali sensitive preparations For blood plasma And infusion fluids. Parenteral Used once only.
Type – III Treated sodalime glass (soda – glass with surface treatment) – sulphured container	SiO– 75% Al_2O_3 – 2% Na_2O – 15% K_2 – O – 0.5% B_2O_3 – 3% CaO – 12%	Moderate Hydrolytic Resistance (Limited Alkalinity) Easy alkali to water Flakes separated out loss of Brilliance Cheap east	Medicine Bottle for solid medicament. Only for non aqueous preparations For providers for injection Not for alkali sensitive products Not for Parenteral used once only
Type – NP (Non – Parenteral)		General purpose soda – lime glass	Used for oral and Topical preparation
Colored glass	In addition to glass composition metal salts carbon and sulphur OR Iron and manganese for Amber color	Does not allow the UV rays to pass through it.	Used for light sensitive Product Do not use for Parenteral unless specified
Lead free glass		As lead monoxide is used in manufacturing of glass cause Lead poisoning	Used for liquid preparations Used when preparation meant for lead poisoning e.g. Na E.D.T.A.
Silicone treated	Surface is treated with dimethyl siloxane	As hydrophobic nature not welled so product do not cling to surface	

Disposable glass syringes have been replaced by plastic syringes. Glass containers for aerosol purpose present risk of breakage as compared to the metal containers. Metal containers on the other side do not offer flexibility in design. Glass is being used for aerosol purpose making use of good bottle strength or adequate bottle strength with an external coating of a flexible plastic (e.g. PVC). Uncoated bottles are used for keeping low pressure aerosols.

Table 4.4 Merits and demerits of Glass.

Merits	Demerits
√ Chemically inert	√ Fragile
√ Impart no odour and taste to the product	√ Less pressure safety and impact Resistance (aerosol containers can withstand pressure below 25 psig and use of less than 50% Propellant is quite safe)
√ Non corrosive	√ High freight cost to its weight
√ Strong and rigid	
√ Impermeable	
√ Transparent and sparkle	
√ FDA approved	
√ No change on aging	

After the advent of plastic as a packaging material, it was once considered that the application of glass would come down drastically. However except for the breakage risk, glass the best product protection with distinct advantages (Table 4.4). This is testified also with the fact that FDA acceptance of glass rarely presents any problem. Lightweight coated glass with improved color range and strength seem to be very promising for pharmaceutical packaging in the future.

In addition to above classification the glass industry classifies glass in various types depending upon its use and nature. The various categories are as follows:

Aluminium oxynitride	Acrylic	Art glass
Aerogel	Beveled glass	Blenko Glass Company
Blown plate	Borosilicate glass	Broad sheet
Bulletproof glass	Calumite	Crystal glass
Cylinder blown sheet	Edinburgh crystal	Fiberglass
Float glass	Fire polishing	Glassblowing
Glass container industry	Glass-reinforced plastic	Glass recycling
Lexan	Machine drawn cylinder sheet	
Magnifying glass	Marvering	Metallic glass
Millefiori	Murano glass	
Obsidian (volcanic glass)		
Opaline glass	Polished plate	Prince Rupert's Drops
Pyrex	Recycling glass	Sea glass
Smart glass	Stained glass	Supercooling
Toughened glass	Trinitite	Tyrone Crystal
Venetian Glass	Waterford Crystal	Water glass

Favrile iridescent glass - Tiffany's technique to make stained glass art

4.3 Glass Properties

Glass properties can be calculated through statistical analysis of glass databases such as SciGlass and Interglad. If the desired glass property is not related to crystallization (e.g., liquidus temperature) or phase separation, linear regression can be applied using common polynomial functions up to the third degree. Below is an example equation of the second degree. The *C*-values are the glass component concentrations like Na_2O or CaO in percent or other fractions, the *b*-values are coefficients, and *n* is the total number of glass components. The glass main component silica (SiO_2) is excluded in the equation below because of over-parameterizations due to the constraint that all components sum up to 100%. Many terms in the equation below can be neglected based on correlation and significance analysis. Further details and examples are available at Glassproperties.com.

$$\text{Glass property} = b_0 + \sum_{i=1}^{n} \left(b_i C_i + \sum_{k=i}^{n} b_{ik} C_i C_k \right)$$

The liquid's temperature has been modeled using neural networks regression in the following article: C. Dreyfus, G. Dreyfus: "A machine learning approach to the estimation of the liquid's temperature of glass-forming oxide blends. It is often required to optimize several glass properties simultaneously, including production costs. This can be performed in a spreadsheet as follows:

1. Listing of the desired properties;
2. Entering of models for the reliable calculation of properties based on the glass composition, including a formula for estimating the production costs;
3. Calculation of the squares of the differences (errors) between desired and calculated properties;
4. Reduction of the sum of square errors using the Solver option in Microsoft Excel with the glass components as variables.

The many advantages that glass offers have enabled it to exist in the pharmaceutical packaging till date.

4.3.1 Chemical Properties

Sodium/Alkali Leaching

All soda-lime type glasses and some borosilicate glasses contain sodium or alkali metal ions. Prolonged exposure to liquids or vapor, such as water, will cause the sodium/alkali ions to migrate to the surface of the glass. This can cause a cloudiness or haze on the surface of the glass. Porous coatings may also incur this phenomenon, causing a disruption of the bond between the coating and the glass surface. In high humidity or critical surface applications, this must be considered when specifying the material. Placing a "barrier" coating, such as silicon dioxide, on the glass will limit the amount of reaction.

Alkali Resistance (AR)

The test occurs with large quantities of a warmed (50°C) alkali solution (pH 12). The time required to remove a layer of 0.1 mm is used to designate four alkali resistance classes AR 1 to AR 4.

Acid Resistance (SR) per ISO 8424:1987

Large quantities of a strong acid (pH 0.3) are applied at 25°C. The time to remove a layer 0.1 mm thick is measured and classified by SR values.

4.3.2 Electrical Properties

Volume Resistivity

It is the resistance in ohms between opposite faces of a centimeter cube of the glass tested. Its value is stated in ohms-centimeters, and it is tested per ASTM D 257. This is important when glass is used as an electrical insulator.

Surface Resistivity

It is the ratio of the potential gradient parallel to the current along its surface, to the current per unit width of the surface. Surface resistivity is measured according to ASTM D 257. This usually specified as ohms per square centimeter (Ω/cm2) or just ohms per square. This method is used to measure the conductivity of coated glasses.

Dielectric Constant

Dielectric constant of a glass is the ratio of the energy stored in a condenser with the glass as the dielectric, compared with the energy stored by the same condenser with air as the dielectric. It is a measure of the ability of a glass to store electrical energy and varies with the frequency of the voltage applied to the condenser. The dielectric strength is measured per ASTM D 150. This is important when the glass is used as a substrate for electrical or electronic devices.

4.3.3 Mechanical Properties

Stress

Stress is defined as the perpendicular force per unit area applied to an object, in a way that compresses (compressive stress) or stretches (tensile stress) the object. Strength is the ability of glass to withstand these stresses. Non-strengthened glass materials have relatively low tensile strength yet high compressive strength. Therefore, most glass breakage is due to tensile stress failure.

Poisson's Ratio

The ratio of the transverse contraction of a bar of glass to the elongation per unit length. Most glasses have a Poisson's ratio of around 0.22.

Young's Modulus

Young's Modulus or Modulus of Elasticity; the amount of stress required to produce a unit change in length (strain). Expressed in pounds per square inch (PSI) or MegaPascals (Mpa).

Modulus of Rupture

The Modulus of Rupture (MOR) strength of glass is determined by tests conducted in accordance with ASTM C 158. This test measures the bending (flexural) strength of glass. Expressed in PSI or Mpa.

Shear Modulus

The amount of force that is required to withstand, by applying the force so that the glass is sheared or twisted.

Bulk Modulus

The inverse of the compressibility of the material, if the material does not exhibit any time dependent response.

Density

It is the mass value per unit of volume, such as grams per cubic centimeter (g/cm^3).

Specific Gravity

It is the ratio of the density of the glass to the density of water, given as a fraction.

Knoop Hardness

A measure of hardness employing a diamond pyramid indenter. The measure of hardness is known as the Knoop hardness number (KHN).

4.3.4 Optical Properties

Refractive Index

It is the ratio of the velocity of propagation of an electromagnetic wave in a vacuum to its velocity in the glass. This determines how much a light wave is "bent" when entering and leaving the surface of the glass. This property is important in producing certain optical devices or effects, such as lenses. The refractive index differs depending on the specific wavelength of light. Therefore, the refractive index is always specified at stated wavelengths, generally corresponding to characteristic material spectral emission lines, such as the sodium "D" line – the most commonly used wavelength value.

Abbe Value

It is the ratio of refractivity to dispersion in an optical medium. Specifically: (nd-1)/(nF-nC), where n is the index of refraction for the spectral lines Fraunhofer d, F and C, respectively.

Dispersion

A measure of the separation of light into its component colors, such as a prism dispersing white light into a color band, or a rainbow effect.

Absorption

The amount of light energy converted to heat within a material that is not transmitted nor reflected. The amount of absorption varies by material type and wavelength. Tinted materials will absorb more light than clear materials.

Transmission

The passage of light through a material; light not reflected back to its source or absorbed is transmitted through the glass. Values are given as a percentage of the original source light energy versus the transmitted light energy. These values vary with wavelength and for each material. For example, soda-lime glass has very low transmission at a wavelength of 265nm but fused silica transmits well at this wavelength.

Reflectivity

It is the return of light from a surface with no change in wavelength. This is generally given as a percentage of the source light energy versus the reflected energy and is dependent upon the angle of incidence (the angle between the light source and the glass surface, usually specified from "normal" – a line perpendicular to the glass surface) and wavelength.

Birefringence Constant

Birefringence is the separation of a beam of light into two beams as it passes through a doubly refracting glass. Differing stress layers in a glass will cause birefringence and therefore birefringence may used to measure the amount of stress in a material. Since the amount of birefringence differs by material, the birefringence constant must be considered when determining stress levels.

4.3.5 Thermal Properties

Coefficient of Thermal Expansion (CTE)

The property of a material to expand in size as the temperature is raised. Most glasses are have a relatively linear expansion rate between 0°and 300° C. This is an important factor to consider when placing the glass in a frame, since glass expands much more than most metals and plastics and may cause breakage upon cooling. Glass such as Borofloat is designed with a CTE to match silicon and when bonded together will expand and contract with minimal stress to the device.

Thermal Conductivity

It is the ability to conduct heat through the glass or away from a heat/light source. This is important when considering glass as a view port for high temperature or high infrared applications.

Maximum Operating Temperature

The maximum temperature recommended for use in an application. This may be a long or short-term rating. Note: for strengthened glass, never exceed a temperature 100° C. below the Strain Point for short term or 200° C. below the Strain Point long term in the application.

Specific Heat

A property that indicates the amount of energy glass stores for each degree increase in temperature, on a per unit mass basis. Its units are J/kg-K.

Surface Emissivity

The relative emissive power of a body compared to that of an ideal blackbody. In other words, the fraction of thermal radiation emitted compared to the amount emitted if the body were a blackbody. By definition, a blackbody has a surface emissivity of 1. The emissivity is also equal to the absorption coefficient, or the fraction of any thermal energy incident on a body that is absorbed.

Working Temperature

This is the temperature at which the glass may be fully re-shaped. The glass may be formed into any shape and will sag completely under its own weight.

Thermal Shock

It is the amount of temperature differential that can be withstood by glass when subjected to a rapid change. Heating the glass to a specific temperature and then exposing to a cooler temperature, usually ice water, verifies the specific rating. Strengthened glass, whether heat-treated or chemically strengthened, significantly improves the rating over annealed glass.

Thermal Gradient

It is defined as the maximum temperature differential between the front surface and the back surface that will cause a 1000 PSI tensile stress on the cooler surface.

Softening Point

It is the temperature at which unsupported glass will begin to sag. If the temperature is reduced slightly, the glass will remain in the sagged shape. For soda-lime glass, this point is 726° C (1340° F.).

When heat treating glass is heated, the softening point should not be exceeded during heating to prevent roller wave or other shape distortion. Generally, heat treat processing equipment targets approximately 1180-1200° F. maximum glass temperature during the glass heating phase.

Annealing Point

This is the temperature at which glass will relieve stresses (either compressive or tensile) in a matter of minutes. For soda-lime glass the annealing point is 546° C. (1015° F.). When heat treating of glass is heated, all portions of the glass must exceed this temperature in order to prevent "cold-cracking" during quench.

When fabricating glass and to prevent warping or bowing due to unequal front and back surface stresses the glass must be properly annealed. To anneal glass, whether during the manufacturing of the glass or after processing of the glass, all portions of the glass must be cooled uniformly from over the annealing point to under the strain point. (see Strain Point below).

Strain Point

This is the temperature at which glass will relieve stresses over a period of hours. For soda-lime glass this temperature is 514° C. (957° F.). Since this is a point of reference only, stresses will continue to relieve over extended periods of time below this temperature. The stress relief is cumulative, so this is a gradual degradation of stress if glass remains in the area of the strain point for extended periods.

Therefore, heat treated glass should not be used at a temperature exceeding 250° C. (482° F.) for short periods of time or 230° C. (428° F.) for extended periods of time (see Maximum Operating Temperature).

Glass is chemically strengthened at temperatures slightly below generally 25° C. (50° F.) below the strain point or else the compressive stresses introduced into the glass by ion-exchange will be relieved as the are produced, resulting in no significant strengthening. It is possible to increase the strength parameters by reducing the temperature at which ion-exchange occurs by more than the 25° C.

Normal Service

No breakage from excessive thermal shock is assumed.

Extreme Limits

The glass will be very vulnerable to thermal shock. Recommendations in this range are based on mechanical stability considerations only. Tests should be made before adopting final designs. These data approximate only. Intermittent only - short periods of time.

Coefficient of Thermal Expansion

In short it is the relative increase in size of a material when heated.

Tensile Strength

It is the resistance of a material to breakage under the stress of pulling or stretching.

Design Tensile Strength

It is the anticipated tensile load that a material can withstand over an indefinite period of time.

Thermal Shock

This property is based on plunging sample into cold water after oven heating. Resistance of 97° means no breakage if heated to 147°F and plunged into water at 50°F. Tempered samples have over twice the resistance of annealed glass.

Temperature Gradient

It is the temperature differential between two surfaces of a tube or constrained plate that will cause a tensile stress of 1000 psi on the cooler surface.

4.4 Glass as a Liquid

Glass is generally treated as an amorphous solid rather than a liquid, though different views can be justified since characterizing glass as either 'solid' or 'liquid' is not an entirely straightforward matter. However, the notion that glass flows to an appreciable extent over extended periods of time is not supported by empirical research or theoretical analysis.

A myth does exist that glass rods and tubes can bend under their own weight over time. To test this, in the 1920s, Robert John Rayleigh, son of the Nobel Prize winner John William Rayleigh, conducted an experiment on a 1 metre (~39 in) long, 5 millimetre (~3/16 in) thick glass rod, which was supported horizontally on two pins with a 300 gram (~0.66 lb) weight in the middle. Apart from the initial bending of 28 millimetre (~1.1 in), the position of the weight did not change until the end of the experiment, which lasted for 7 years. At the same time, another man, a worker of General Electric named K. D. Spenser, conducted a similar experiment independently. Two months after Rayleigh, he published his own results which also disproved the myth. Spenser suggested that the myth was composed before the 1920s, when the tubes were made by hand, and naturally some of them were curved to begin with. Over time the straight tubes were taken away, and only the curved ones remained. Some people probably thought it was the glass flowing.

Some people believe glass is a liquid due to its lack of a first-order phase transition. There is no discontinuous change of density and no latent heat of fusion. However, glass does go through a second-order phase transition where the thermal expansivity and heat capacity change markedly.

Although glass has properties of a super cooled liquid, it is generally classed as solid at room temperature. There is also the problem that a super cooled liquid is still a liquid — moves and behaves like a liquid, not a solid — but is below the freezing point of the material and will crystallize almost instantly if a crystal is added as a core.

In addition to the quartz, sodium carbonate, and calcium carbonate traditionally used in glassmaking, boron is used in the manufacture of borosilicate glass. Typically, the resulting glass composition is about 70% silica, 10% boric oxide, 8% sodium oxide, 8% potassium oxide, and 1% calcium oxide. Though somewhat more difficult to make than traditional glass (Corning conducted a major revamp of their operations to make it), it is economical to produce because its superior durability, chemical and heat resistance finds excellent use in chemical laboratory equipment, cookware, lighting, and in certain cases, windows.

4.5 Colored Glasses

Amber is the fossilized resin of ancient pine trees, submerged under the sea in thin veins. It can be gathered along the North Sea coasts of East Anglia, south-west Jutland and the southern shores of the Baltic. It is washed loose onto beaches from its deposits by sea currents, causing it to float to the surface, especially during violent storms. Ranging in colour from a dark, reddish brown to a translucent straw, it was a treasured material, particularly by the Vikings. Other uses for amber coloured glass, apart from bead making, includes pendants, amulets, gaming pieces, spindle-whorls, and finger rings.

There were two main ways of making glass: either from the raw materials of quartz and soda (or by the late tenth century, quartz and potash), or more usually by melting down broken glass (cullet) and then re-using it.

It goes without saying that the broken glass or cullet used had to be of the required colour or clear, so that no expensive new minerals were added to colour it. The problem of getting enough of one colour was overcome by importing blocks of coloured glass taken from continental mosaics (tesserae) and windows (so there's nothing new in recycling glass).

When glass is made in this fashion, it is clear or has a slight green tinge. In order to colour it minerals were added; copper for red, blue or green, iron for black, tin for yellow. Coloured glass found includes pale blue, dark blue, blue-green, emerald green, olive green, amber, yellow-brown, red and black.

Glass supplies literally thousands of colors and finishes. Lighter colors such as pastels and off-whites are affected by the natural greenness of clear glass. As a result, Glass usually recommends the use of low-iron glass such as Starphire or OptiWhite, which has very little color content to the glass.

Glass can be changed into various colors. That depends on how the glass was mounted. Most installations are done with an adhesive to firmly attach glass to a wall and cannot be removed without the glass breaking (similar to removing one large tile). However, if the glass was installed with mechanical fasteners, it can be easily changed to another piece of glass with a different color applied. But in no event can the actual Glass coating be removed from glass. It bonds chemically to the glass; once it is applied; it doesn't come off.

The color of a transparent material such as glass is due to the absorption of light at certain wavelengths in the visible spectrum (350-700 nm). There are several types of absorption: - Electron transfer between cations and anions. Depending on the ions used, absorption can occur in either the visible or non-visible spectrum. Electron transfer between Fe^{III+} and O^{II-} corresponds to ultraviolet absorption. The replacement of O^{II-} by S^{II-} moves the absorption band into the visible range, towards 410-420 nm. This produces amber colored glass on which are based most of the current colors used in bottles.

- Electron transfer from one electronic layer to another within the same cation. The wavelength of the absorption depends on the cation and the glass matrix. For instance Fe^{II} absorbs around 1000-1100 nm and therefore gives the glass a blue tint.
- The interaction of light with specific particles blended within the glass: this produces for example copper ruby red in which small aggregates of copper Cu^{I} and Cu^{0} can be formed by heat treatment. This is also the case for silver-yellow or gold-red.
- The interaction of light with small crystals formed during cooling: this renders the glass white as a result of light diffusion.

CHAPTER 5

Metals in Packaging

Summary

5.1 Introduction

Metal containers being used in the pharmaceutical industry include small – elongated collapsible tubes and shallow drums. Metal containers have dominated themselves in the field of aerosols. Though glass, plastic and plastic coated glass aerosols are finding their own specialized application, metal aerosols are likely to retain the bulk of the market as long as cost advantages are offered. Aluminium foil is another metal extensively used in the pharmaceutical industry.

5.2 Aluminium

Aluminum is too reactive chemically to occur in nature as the free metal. Instead, it is found combined in over 270 different minerals. The chief source of aluminium is bauxite ore. Aluminium is remarkable for its ability to resist corrosion (due to the phenomenon of passivation) and its light weight.

5.3 Aluminium Foil

Aluminium foil is sometimes known as al-foil or alu-foil. It is also often called tinfoil, although it is not made from tin, or as silver paper although it is not made from silver; or in North America, as Reynolds wrap after Reynolds Metals, the leading manufacturer when it was introduced on the American market (much to the chagrin of Alcoa, Reynolds' main competitor, which had its brand "Alcoa Wrap" referred to as "Alcoa Reynold's Wrap").

5.4 Tin

Tin acts as a catalyst when oxygen is in solution and helps accelerate chemical attack. Tin forms the dioxide SnO_2 when it is heated in the presence of air. SnO_2, in turn, is feebly acidic and forms stannate (SnO_3^{-2}) salts with basic oxides. Tin can be highly polished and is used as a protective coat for other metals in order to prevent corrosion or other chemical action.

5.5 Lead

Lead has been commonly used for thousands of years because it is widespread, easy to extract and easy to work with. It is highly malleable and ductile as well as easy to smelt. In the early Bronze Age lead was used with antimony and arsenic. Lead is very toxic, and lead poisoning was recognized even by the ancients. Similarly, in the Twentieth Century, the use of lead in paint pigments was sharply reduced because of the danger of lead poisoning, especially to children.

5.6 Collapsible Metal Tube

The collapsible metal tube is an attractive container that permits controlled amounts to be dispensed easily, with good reclosure, and adequate protection of the product. The risk of contamination of the portion remaining in the tube is minimal because the tube does not 'suck back'.

Any ductile metal that can be worked cold is suitable for collapsible tubes, but the most common ones in use are tin (15%), aluminium (60%), and lead (25%). Tin is the most expensive, and lead is the cheapest.

5.1 Introduction

Metal containers are strong, relatively unbreakable, opaque, and impervious to water vapour, gases odour and bacteria. They are resistant to high and low temperatures. Due to their chemical reactivity, metals require the application of coating and lacquers to prevent chemical reaction and corrosion. Now a days special coatings and coating techniques are being developed for this purpose. Metal containers being used in the pharmaceutical industry include small-elongated collapsible tubes and shallow drums. Metal containers have dominated themselves in the field of aerosols. Though glass, plastic and plastic coated glass aerosols are finding their own specialized application, metal aerosols are likely to retain the bulk of the market as long as cost advantages are offered. Aluminium foil is another metal extensively used in the pharmaceutical industry.

Metals are also used for the formation of closures. Some of these closures are similar to those used on glass and plastic containers e.g. plastic and metal screw closures and friction closures such as plug or slip lids. Other metal closures include lever lids and permanent mechanically seamed-on closures.

Tin plate was once used for pastille tins, ointments tins and various build-up containers for powders, tablets, capsules, etc. These were followed by the use of Aluminium. The use of metals is reducing, as they cannot compete with the cost of plastic and glass. Some metals currently in used in the pharmaceutical packaging include tin plate, tin-free steel, Aluminium, alloy of Aluminium and stainless steel. Aluminium and its alloys are used for aerosol containers, rigid and collapsible tubes, shallow drawn containers, blisters and sachet packaging. Tin plate containers are used for the formation of built-up containers and shallow drawn containers. Stainless steel is widely used in pharmaceuticals for mixing vessels and manufacturing equipment. It is also used for packaging of high-pressure aerosols. Tin is occasionally used for collapsible tubes. Metal containers still play a very important role in packaging, accounting for one-quarter of the total sales of all packaging materials.

5.2 Aluminium

Aluminium is the most abundant of all metals and the third most abundant element in the Earth's crust, after oxygen and silicon. It makes up about 8% by weight of the Earth's solid surface.

Aluminum is too reactive chemically to occur in nature as the free metal. Instead, it is found combined in over 270 different minerals . The chief source of aluminium is bauxite ore. Aluminium is remarkable for its ability to resist corrosion (due to the phenomenon of passivation) and its light weight. Structural components made from aluminium and its alloys are vital to the aerospace industry and very important in other areas of transportation and building.

Properties

Aluminium is a soft, lightweight metal with appearance ranging from silvery to dull gray, depending on the surface roughness. Aluminium is nontoxic, nonmagnetic, and nonsparking. The yield strength of pure aluminium is 7–11 MPa, while aluminium alloys have yield strengths ranging from 200 MPa to 600 MPa. Aluminium has about one-third the density and stiffness of steel. It is ductile, and easily machined, cast, and extruded.

Corrosion resistance is excellent due to a thin surface layer of aluminium oxide that forms when the metal is exposed to air, effectively preventing further oxidation. The strongest aluminium alloys are less corrosion resistant due to galvanic reactions with alloyed copper.

Properties of Aluminium

Name, symbol, number	Aluminium, Al, 13
Chemical series	poor metals
Standard atomic weight	26.9815386(8) g·mol−1
Phase	solid
Melting point	933.47 K (660.32 °C, 1220.58 °F)
Boiling point	2792 K (2519 °C, 4566 °F)
Density (near r.t.)	2.70 g·cm−3
Liquid density at m.p.	2.375 g·cm−3
Heat capacity	(25 °C) 24.200 J·mol−1·K−1
Heat of fusion	10.71 kJ·mol−1
Heat of vaporization	294.0 kJ·mol−1
Thermal conductivity	(300 K) 237 W·m−1·K−1
Electrical resistivity	(20 °C) 26.50 nΩ·m
Young's modulus	70 GPa
Shear modulus	26 GPa
Bulk modulus	76 GPa
Poisson ratio	0.35
Vickers hardness	167 Mpa
Mohs hardness	2.75

Aluminium atoms are arranged in an FCC structure. Aluminium has a high stacking-fault energy of approximately 200 mJ/m^2.

Aluminium is one of the few metals which retain full silvery reflectance in finely powdered form, making it an important component of silver paints. Aluminium mirror finish has the highest reflectance of any metal in the 200–400 nm (UV) and the 3000–10000 nm (far IR) regions, while in the 400–700 nm visible range it is slightly outdone by silver and in the 700–3000 (near IR) by silver, gold, and copper

Aluminium is a good thermal and electrical conductor, by weight better than copper. Aluminium is capable of being a superconductor, with a superconducting critical temperature of 1.2 Kelvin.

General use

Aluminium is the most widely used non-ferrous metal. Global production of aluminium in 2005 was 31.9 million tonnes. It exceeded that of any other metal except iron (837.5 million tonnes).

Relatively pure aluminium is encountered only when corrosion resistance and/or workability is more important than strength or hardness. Pure aluminium serves as an excellent reflector (approximately 99%) of visible light and a good reflector (approximately 95%) of infrared. A thin layer of aluminium can be deposited onto a flat surface by chemical vapour deposition or chemical means to form optical coatings and mirrors.

Pure aluminium has a low tensile strength, but when combined with thermo-mechanical processing, aluminium alloys display a marked improvement in mechanical properties, especially when tempered. Aluminium alloys form vital components of aircraft and rockets as a result of their high strength-to-weight ratio. Aluminium readily forms alloys with many elements such as copper, zinc, magnesium, manganese and silicon (e.g., duralumin). Today, almost all bulk metal materials that are referred to loosely as "aluminium," are actually alloys. For example, the common aluminium foils are alloys of 92% to 99% aluminium.

Some of the common uses for aluminium metal are in:

- Transportation (automobiles, aircraft, trucks, railway cars, marine vessels, bicycles etc.)
- Packaging (cans, foil, etc.specially for packaging of medicines, cosmetics and foodstuff)
- Water treatment
- Electrical transmission lines for power distribution
- MKM steel and Alnico magnets

- Super purity aluminium (SPA, 99.980% to 99.999% Al), used in electronics and CDs.
- Treatment against fish parasites such as *Gyrodactylus salaris*.
- Construction (windows, doors, siding, building wire, etc.)
- Cooking utensils
- Heat sinks for electronic appliances such as transistors and CPUs.
- Powdered aluminium is used in paint, and in pyrotechnics such as solid rocket fuels and thermite.

Precautions

Aluminium is a neurotoxin that alters the function of the blood-brain barrier. It is one of the few abundant elements that appears to have no beneficial function to living cells. A small percent of people are allergic to it — they experience contact dermatitis from any form of it: an itchy rash from using styptic or antiperspirant products, digestive disorders and inability to absorb nutrients from eating food cooked in aluminium pans, and vomiting and other symptoms of poisoning from ingesting such products as Amphojel, and Maalox (antacids). In other people, aluminium is not considered as toxic as heavy metals, but there is evidence of some toxicity if it is consumed in excessive amounts. The use of aluminium cookware, popular because of its corrosion resistance and good heat conduction, has not been shown to lead to aluminium toxicity in general. Excessive consumption of antacids containing aluminium compounds and excessive use of aluminium-containing antiperspirants are more likely causes of toxicity. In research published in the Journal of Applied Toxicology, Dr. Philippa D. Darby of the University of Reading has shown that aluminium salts increase estrogen-related gene expression in human breast cancer cells grown in the laboratory. These salts' estrogen-like effects have lead to their classification as a metalloestrogen.

It has been suggested that aluminium is a cause of Alzheimer's disease, as some brain plaques have been found to contain the metal. Research in this area has been inconclusive; aluminium accumulation may be a consequence of the Alzheimer's damage, not the cause. In any event, if there is any toxicity of aluminium it must be via a very specific mechanism, since total human exposure to the element in the form of naturally occurring clay in soil and dust is enormously large over a lifetime.

Mercury applied to the surface of an aluminium alloy can damage the protective oxide surface film by forming amalgam. This may cause further corrosion and weakening of the structure. For this reason, mercury thermometers are not allowed on many airliners, as aluminium is used in many aircraft structures.

Powdered aluminium can react with Fe_2O_3 to form Fe and Al_2O_3. This mixture is known as thermite, which burns with a high energy output. Thermite can be produced inadvertently during grinding operations, but the high ignition temperature makes incidents unlikely in most workshop environments.

5.3 Aluminium Foil

Large roll of Aluminium foil, with micrometer showing a thickness of 0.0005 inches (0.013mm). Aluminium foil is aluminium prepared in thin sheets (about 0.02 mm in thickness). As a result of this, the foil is extremely pliable, and can be bent or wrapped around objects with ease.

Aluminium foil is sometimes known as **al-foil** or **alu-foil**. It is also often called **tinfoil**, although it is not made from tin, or as **silver paper** although it is not made from silver; or in North America, as **Reynolds wrap** after Reynolds Metals, the leading manufacturer when it was introduced on the American market (much to the chagrin of Alcoa, Reynolds' main competitor, which had its brand "Alcoa Wrap" referred to as "Alcoa Reynold's Wrap").

Millions of tons of aluminium foil are used throughout the world in the protection and packaging of foods, cosmetics and chemical products. Usually, an extremely thin layer (0.0065 mm or 6.5 μm) is laminated to other materials, plastics and paper.

Foil made from a thin leaf of tin was commercially available before the aluminium counterpart. In the late 19th century and early 20th century, tin foil was in common use, and some people continue to refer to the new product by the name of the old one. Tin foil is stiffer than aluminium foil. It tends to give a slight tin taste to food wrapped in it, which is one major reason it has largely been supplanted by aluminium and other materials for wrapping food.

The first audio recordings on phonograph cylinders were made on tin foil.

Tin was first replaced by aluminium in 1910, when the first aluminium foil rolling plant, "Dr. Lauber, Neher & Cie., Emmishofen." was opened in Kreuzlingen, Switzerland.

The plant, owned by J.G. Neher & Sons (aluminium manufacturers) started in 1886 in Schaffhausen, Switzerland, at the foot of the Rhine Falls - capturing the falls' energy to produce aluminium. Neher's sons together with Dr. Lauber discovered the endless rolling process and the use of aluminium foil as a protective barrier. From there began the wide use of aluminium foil in the packaging of chocolate bars and tobacco products. Processes evolved over time to include the use of print, colour, lacquer, laminate and the embossing of the aluminium.

An Aluminium foil is the material most widely used for packaging of tablets and capsules though it is the most expensive constituent of the laminate. Aluminium foils is made from the metal that has 99% purity and the gauge of the foil may vary from 0.006 to 0.040 mm. Aluminium foil offers excellent barrier properties to moisture, gases and light, gives the pack brightness and an attractive look. The foil provides hygienic, tamper evident pack, which is odorless, tasteless and non – toxic. It provides a convenient, safe and versatile packaging for tablets, capsules, liquids and powders.

Aluminium foil is strong and can be laminated. It can form push – through lids of preformed trays for tablets. When used as collapsible tube these do not draw air into them (forming dead ends), which may contaminate the product.

Aluminium foil is particularly suitable for blister packs. The tray of the blister may be made of PVC/ PVdC while the lidding is done with the help of Aluminium foil. Aluminium foil can be made into alloy, which can alter its properties like brittleness, toughness and ductility. Aluminium foil is also used for forming strips packs. Strip packs provide economical high barrier performance pack and can be filled at high speed on modern, sterile filling lines. In case of sachets and pouches Aluminium packs are robust and keep the product in good condition. These can withstand rigorous "burst" tests and are convenient and economical.

Aluminium foil laminated with paper or plastic is frequently used as a heat – sealed membrane hermetically closing the container under a plastic screw cap. The resulting Aluminium membrane provides excellent barrier properties preventing moisture or gas transmission and tamper evident seal.

5.4 Tin

Tin is a malleable, ductile, highly crystalline, silvery-white metal; when a bar of tin is bent, a strange crackling sound known as the "tin cry" can be heard due to the breaking of the crystals. This metal resists corrosion from distilled, sea and soft tap water, but can be attacked by strong acids, alkalis, and by acid salts.

Description

Boiling point	2875K (2602°C, 4716°F)
Melting point	505.08 K (231.93°C, 449.47°F)
Density (near r.t.)	5.769 g·cm^{-3}
Liquid density at m.p.	6.99 g·cm^{-3}
Young's modulus	50 GPa
Bulk modulus	58 GPa
Poisson ratio	0.36
Mohs hardness	1.5
Brinell hardness	51 MPa

Tin acts as a catalyst when oxygen is in solution and helps accelerate chemical attack. Tin forms the dioxide SnO_2 when it is heated in the presence of air. SnO_2, in turn, is feebly acidic and forms stannate (SnO_3^{-2}) salts with basic oxides. Tin can be highly polished and is used as a protective coat for other metals in order to prevent corrosion or other chemical action. This metal combines directly with chlorine and oxygen and displaces hydrogen from dilute acids. Tin is malleable at ordinary temperatures but is brittle when it is heated.

Tin bonds readily to iron, and has been used for coating lead or zinc and steel to prevent corrosion. Tin-plated steel containers are widely used for food preservation, and this forms a large part of the market for metallic tin. Speakers of British English call them "tins"; Americans call them "cans" or "tin cans". One thus-derived use of the slang term "tinnie" or "tinny" means "can of beer". The tin whistle is so called because it was first mass-produced in tin-plated steel.

Tin foil was once a common wrapping material for foods and drugs; replaced in the early 20th century by the use of aluminium foil, which is now commonly referred to as *tin foil.* Hence one use of the slang term "tinnie" or "tinny" for a small retail package of a drug such as cannabis or for a can of beer.

Elemental tin is an essential nutrient, needed in very small amounts. The small amount of tin that is found in canned foods is not harmful to humans.

Certain organic tin compounds, organotin, such as triorganotins (see tributyltin oxide) are toxic and are used as industrial fungicides and bactericides.

5.5 Lead

Lead has a dull luster and is a dense, ductile, very soft, highly malleable, bluish-white metal that has poor electrical conductivity. This true metal is highly resistant to corrosion. Because of this property, it is used to contain corrosive liquids (e.g. sulfuric acid). Lead can be toughened by adding a small amount of antimony or other metals to it. Lead is the only metal in which there is zero Thomson effect. Lead is also poisonous. All lead, except ^{204}Pb, is the end product of a complex radioactive decay (see isotopes of lead below).

Description

Boiling point	2022K(1749°C, 3180°F)
Melting point	600.61K(327.46°C, 621.43°F)
Density (near r.t.)	11.34 $g \cdot cm^{-3}$
Liquid density at m.p.	10.66 $g \cdot cm^{-3}$
Heat of fusion	4.77 $kJ \cdot mol^{-1}$
Thermal conductivity	(300□K) 35.3 $W \cdot m^{-1} \cdot K^{-1}$
Young's modulus	16 GPa
Shear modulus	5.6 GPa
Bulk modulus	46 GPa
Poisson ratio	0.44
Mohs hardness	1.5
Brinell hardness	38.3 MPa

Lead has a dull luster and is a dense, ductile, very soft, highly malleable, bluish-white metal that has poor electrical conductivity. This true metal is highly resistant to corrosion. Because of this property, it is used to contain corrosive liquids (e.g. sulfuric acid). Lead can be toughened by adding a small amount of antimony or other metals to it. Lead is the only metal in which there is zero Thomson effect. Lead is also poisonous. All lead, except ^{204}Pb, is the end product of a complex radioactive decay (see isotopes of lead below).

Lead has been commonly used for thousands of years because it is widespread, easy to extract and easy to work with. It is highly malleable and ductile as well as easy to smelt. In the early Bronze Age lead was used with antimony and arsenic. Lead is very toxic, and lead poisoning was recognized even by the ancients. Similarly, in the Twentieth Century, the use of lead in paint pigments was sharply reduced because of the danger of lead poisoning, especially to children.

Lead glass is composed of 12-28% lead oxide. It changes the optical characteristics of the glass and reduces the transmission of radiation.

5.6 Collapsible Metal Tube

The collapsible metal tube is an attractive container that permits controlled amounts to be dispensed easily, with good reclosure, and adequate protection of the product. The risk of contamination of the portion remaining in the tube is minimal because the tube does not 'suck back'.

Any ductile metal that can be worked cold is suitable for collapsible tubes, but the most common ones in use are tin (15%), aluminium (60%), and lead (25%). Tin is the most expensive, and lead is the cheapest. Since tin is the most ductile of these materials, small tubes are often made mostly of tin even though the metal cost is higher. Laminates of tin coated lead provided the appearance and oxidation resistance of tin at lower prices.

The tin that is used for this purpose is alloyed with about 0.5 % copper for stiffening. When lead is used, about 3% antimony is added to increase hardness. The aluminium work hardens when it is formed into a tube, and must be annealed to give it a necessary pliability. Aluminium also hardens in use, sometimes causing tubes to develop leaks.

Tin Containers are preferred for foods, pharmaceuticals, or any product for which the purity is the paramount consideration. Tin is the most chemically inert of all collapsible tube metals. It offers a good appearance and compatibility with a wide range of products.

Aluminium tubes offer significant savings in product shipping costs because of their light weight. They provide the attractiveness of tin at somewhat lower cost.

Lead has lowest cost of all the tube metals and is widely used for nonfood products such as adhesives, inks, paints, and lubricants. Lead should never be used alone for anything taken internally because of the risk of lead poisoning. With internal linings lead tubes are used for such products as fluoride toothpaste. If the product is not compatible with bare metal, the interior can be flushed with wax type formulations or with resin solutions, although the resins and lacquerers are usually spread on.

Wax linings are most often used with water based products in tin tubes, and phenolics, epoxides, and vinyls are used with aluminium tubes, giving better protection than wax, but at the higher cost. Phenolics are most effective with acid products, epoxides protect better against alkaline materials.

CHAPTER 6

Blister and Strip Packages

Summary

6.1 Introduction

Blister packs were developed by George Thonis of the Lakso Company in the 1970s. Blister pack is a term for several types of pre-formed clear plastic packaging used for small consumer products. The two primary components of a blister pack are a pre-formed plastic blister and a printed paperboard card which has a heat-seal coating. Some packs use a foil, plastic, or laminated material to bond to the pre-formed blister.

Strip packs consist of one or two plies, made from regenerated cellulose, Paper, Plastic, foil or a combination of these. The dosage form is inserted into a pocket area against a recess in a heated platen (or roller) or may be inserted in preformed pockets. This from of packaging is used for packaging of tablets and capsules. Two sheets of film are used for making a strip package and the sheets are fed through heated crimping roller. A continuous strip of pockets are formed and cut as required. The product is sealed between the two sheets of film along with a seal around each tablet/ capsule and perforations for easy cutting.

6.2 Blister Packaging

Blister packs are created by means of a form-fill-seal process. A form-fill-seal process means that the blister pack is created from rolls of flat sheet or film, filled with the pharmaceutical product and closed (sealed) on the same equipment. Such equipment is called a blisterline. There are 2 types of forming the cavity into a base web sheet: thermoforming and cold forming. The distribution of the wall thickness achieved by the molding operation depends on the process: whether the material is formed into or over a mould, film thickness, depth of draw, softening temperature, type of plastic, etc. For an identical blister shape, pressure plus plug assistance generally gives the most uniform blister.

*** Thermoforming** *** Cold forming**

6.3 Strip Packaging

Strip packs consist of one or two plies, made from regenerated cellulose, paper, plastic, foil or any combination of these, whereby an item is inserted into a pocket area against a recess in a heated platen or roller. However, some systems now involved the use of performed pockets the by making it increasingly difficult to differentiate between blister and strip packaging.

6.4 Leakage Test for Blister and Strips

It is mandatory to both blister and strip packs to carry out the check for pack integrity. It is of vital importance, both to guarantee product shelf life and to ensure that the product has not become 'contaminated' from some environmental sources.

***Destructive type testing:** ***Non-Destructive type testing**

Pinholes and Package integrity

Pinholes normally refer to the minute holes, which may be present in the foil. Usually the foil of 0.017 mm caliper and above is recognized as commercially pinhole free. The foil with 0.025 mm can normally be 'guaranteed', Pinhole free. Foil below 0.017 mm gradually shows an increasing number of pinholes

6.5 Newer Approaches in Blister and Strip Packaging

(i) ***Innovations in Machine improvements:*** Though for conventional blisters, maximum speeds of over 6,000 items /minute are still relatively rare, much higher speeds are predicted. Recently many machine improvements features have been taken into consideration

(ii) ***Innovations in Materials:*** There occurs perforation problems with 40μm foils. Hence 45-50μm foils have been substituted with gauges available up to 60μm.

(iii) ***Innovations in Extreme barrier packaging:*** Each of the traditional forming material technologies (CFF, PCTFE, COC and PVdC) is available in new forms that address the growing demand for extreme barrier properties.

6.1 Introduction

Blister packs were developed by George Thonis of the Lakso Company in the 1970s. Blister pack is a term for several types of pre-formed clear plastic packaging used for small consumer products. The two primary components of a blister pack are a pre-formed plastic blister and a printed paperboard card which has a heat-seal coating. Some packs use a foil, plastic, or laminated material to bond to the pre-formed blister. In other wards blister packs made of a thermoformed plastic tray with a lidding material made from plastic, paper foil or a combination of these, with a lidding being removed either by pushing through the lid or via a peelable lidding. Today this definition has been extended by cold forming of plastic/ foil combinations and the so-called tropical lidding over the blister.

Blister packs are commonly used as unit dose packaging for Pharmaceutical dosage forms such as tablets, lozenges, soft and hard gelatin capsules, drops etc. Beside Pharmaceuticals, blister packaging is also used in the toy, hardware, cosmetic, novelties, pharmaceutical, confectionary, and food products industries, as well as in many household products. Blister packs can provide barrier protection for self life requirements, and a degree of protection where product (tempered resistant) is a consideration. In the USA, blister packs are mainly used for packing physician samples of drug products or for the sale of Over The Counter (OTC) products in the pharmacy. In other parts of the world, blister packs are the main packaging type since pharmacy dispensing and re-packaging are not common. A series of blister cavities is sometimes called a 'blister card' or 'blister strip' as well as 'blister pack'. In some parts of the world the blister pack is known as a Push-Through-Pack (PTP). Main advantages of unit-dose blister packs over other methods of packing pharmaceutical products are the assurance of product/packaging integrity (including shelflife) of each individual dose and the possibility to create a compliance pack or calendar pack by printing the days of the week above each dose.

Strip-Pack is an innovative packaging system date from the late 1920s (starting with the single Aspro strip in waxed paper). Strip-Pack is an attractive retail display option, which requires minimal shelf space. Strip packs consist of one or two plies, made from regenerated cellulose, paper, plastic, foil or any combination of these, whereby an item is inserted into a pocket area against a recess in a heated platen or roller. However, some systems now involved the use of performed pockets the by making it increasingly difficult to differentiate between blister and strip packaging.

6.2 Blister Packaging

Blister packs are created by means of a form-fill-seal process. A form-fill-seal process means that the blister pack is created from rolls of flat sheet or film, filled with the pharmaceutical product and closed (sealed) on the same equipment. Such equipment is called a blisterline. There are 2 types of forming the cavity into a base web sheet: thermoforming and cold forming.

The principal components of blister packs comprise of two:

1. ***Forming films:*** a formed base web creating the cavity inside which the product fits and
2. ***Lidding material:*** the lidding foil for sticking over the back of blister to ensure air tight packaging of the product.

Forming films

The essential components of pharmaceutical blister packages are forming films (80 – 85% of weight), lidding material (15 - 20% of weight), heat seal coating and printing ink. The blister package is formed by a series of steps. A thermoplastic polymer is softened by application of heat. Vacuum drawing the softened polymer follows this into a contoured mould. The polymer sheet is released from the mould after cooling. This then proceeds to the filling station of the packaging machine. At the filling station, the semi – rigid blister is filled with the product and lidded with heat sealable material. This material may be push through or peelable type. Majority of packs these days are push-through types. In this instance hard foil of 0.015 mm and above is used. This is adequate as the permeability of the plastic tray trends to be far higher than the foil. Opaque of tined materials are necessary if the pack is to meet the UK child-resistant requirements. Opaque films (usually incorporating titanium dioxide) tend to be slightly more permeable than clear material. Permeation may increase with the filler or modifier content of the plastic.

The forming films used for blister packs include.

- Rigid PVC films.
- Coated PVC
- Polystyrene
- CTFE homopolymer
- Cyclo olefin copolymers(COC)
- Polypropylene
- PET

Rigid PVC is a PVC without softening agents. It is clear, has a high flexural strength, good chemical resistance, and low permeability. As such it dominates the market share being economical. PVdC coating reduce the moisture permeability of PVC blister package by a factor of 5 – 10 and hence is very useful for moisture sensitive drugs / combinations. Polypropylene has a low water vapour permeability and easy recyclability. So it was thought of being a good substitute for PVC. But during blistering operations PP requires a very precise temperature control during thermoforming, cooling wrapping and also encounters post processing shrinkage. It is difficult to run PP on the blister machine and so is not a polymer of choice for the purpose. PET and polystyrene are not much is use due to their high vapor permeability. Chlorotrifluoroethylene (CTFE homopolymer) thermoforms easily, has high moisture barrier and exhibits crystal clarity. It is more stable to ageing when compared to PVdC but its high cost makes its use disadvantageous. COC also has high clarity and stiffness and is compatible with thermoforming. Its use combination with layers of PVC, PE, PP, but this increases the cost of packaging.

Lidding material

The most popular caliper or thickness of lidding material range from 0.46-0.61-mm. Lidding material provides the base or main structural component upon which the final blister package is built. These materials should have good printability and be compatible with heat seal coating process. Generally clay coating is added to the lidding material to improve printing. Some of the lidding materials commonly used in pharmaceutical packaging are:

- **Paper-aluminium** combinations are used for child-resistant push-through packages in Europe so the aluminium foil in this case is relatively this. While when the same combination is used in US the aluminium foil is relatively thick to provide for an effective peeling.
- **Paper/PET/ aluminium** laminate is used for peel-off-push-through foil. This laminate is meant to first peel off the paper/ PET laminate from the aluminium and then to push the tablet through the aluminium.
- **Hard aluminium** is the most widely used push-through lidding material. One side of the hard aluminum is coated with print primer while the other side is coated with heat-sealing coating.
- **Soft aluminium** is used for child-resistant push-through foils. This material is also supplied with perforated sealed seams to inhibit easy peel-off.

In case of peelable blisters, heat sealable lacquers have to be 'compatible' with the tray material. If PVC/PVdC is formed with the PVdC on the inside of the blister, a PVdC type compatible lacquer is necessary for the lidding material. All lacquers should be checked for peel strength both initially and at each shelf- life period (1,3,6 moths, etc.) as ageing effects have been detected, e.g. push- through becoming peelable and peelable becoming permanent.

In pharmaceutical industry blister pack is the best form of unit dose packages. It is associated with the ease of dispensation and tamper evidence. For the peeling off kind of backing material, usually a lamination of polyester, PVC (polyvinyl chloride) or its combination with polystyrene and polypropylene is used. Depending upon the need for better moisture protection, laminations with lower water vapour transmission rate (WVTR) are preferred. Such lamination is usually PVC with polyvinylidene chloride.

A novel blister pack has been developed for moisture sensitive drugs in which each blister-containing tablet is connected to a bigger blister-containing desiccant through channels.

Principles in Blister packaging

In blisters a plastic film can be softened by heat then it is formed in a mould utilizing Blister machine. The following procedures are generally utilized:

1. ***Mechanical method:*** in this the forming occurs between male and female moulds.
2. ***Negative pressure (Vacuum):*** It draws the softened film over/ into a mould
3. ***Compressed pressure:*** In this composed air forces are utilized to distribute film over or into a mould.
4. ***Multiple Combinations:*** Two or more of above described methods are used to make blister.

Utilizing any of the methods, the blister trays are formed which are then filled, lidded and cut out into specific sizes and configurations.

Factors affecting blister thickness

The following factors mostly governs the wall thickness achieved by the molding operation:

- ***Material :*** The nature of material e.g. softening temperature, hardening time, elasticity etc.

- ***Process:*** Pressure and plug system usually produce most uniform blisters. Compressed pressure forming produces thicker top section while in vacuum forming, it is thicker near the base and thinner at the top.
- ***Other factors:*** They include depth of draw, softening temperature, thickness of film, etc.

Essential machine operations

The fundamental machine operations performed on a machine includes:

1. Heating and thermoforming
2. Cold forming.
3. Feeding and Filling
4. Heat sealing of the lidding material
5. Punching out or guillotining the tray from the web.
6. Printing of the lidding web,
7. Batch marking,
8. Product check/rejection
9. Perforating or scoring between blister,
10. Code reading and disposal of trim (which –up shredding).

Thermoforming

In the case of thermoforming, a plastic film or sheet is unwound from the reel and guided though a pre-heating station on the blister line. The temperature of the pre-heating plates (upper and lower plates) is such that the plastic will soften and become moldable. The warm plastic will then arrive in a forming station where a large pressure (4 to 8 bar) will form the blister cavity into a negative mold. The mold is cooled such that the plastic becomes rigid again and maintains its shape when removed from the mold. In case of difficult shapes, the warm film will be physically pushed down partially into the cavity by a "plug-assist" feature.

Either infrared heaters or contact heaters are used for heating of the reel–fed base (tray) web. Heat may be applied at the forming stage or at a preheat station with or without further heating. The preheat station may be either a cylinder of platens; in certain instances the platens may be differentially heated (top and bottom). The moulds into which the plastic if formed can be cooled by air, water or chilled water. The web may be held grippers, rollers, etc. as it is passed through the machine. Platen machines usually impart less tension to the web but generally operate at lower speeds i.e. 8-12 m/min whereas 12-16 m/min can be achieved by continuous motion or by reciprocating motion machine with platens. Commonly only the simpler materials can be thermoformed on cylinders using vacuum (e.g. PVC, polystyrene, PVDC coated PVC) (Fig. 6.1). Platens using pressure forming particularly with plug assistance, not only offer more uniform but also can utilize the more complex materials, e.g. Alum/PVC, polypropylene, co-extrusions.

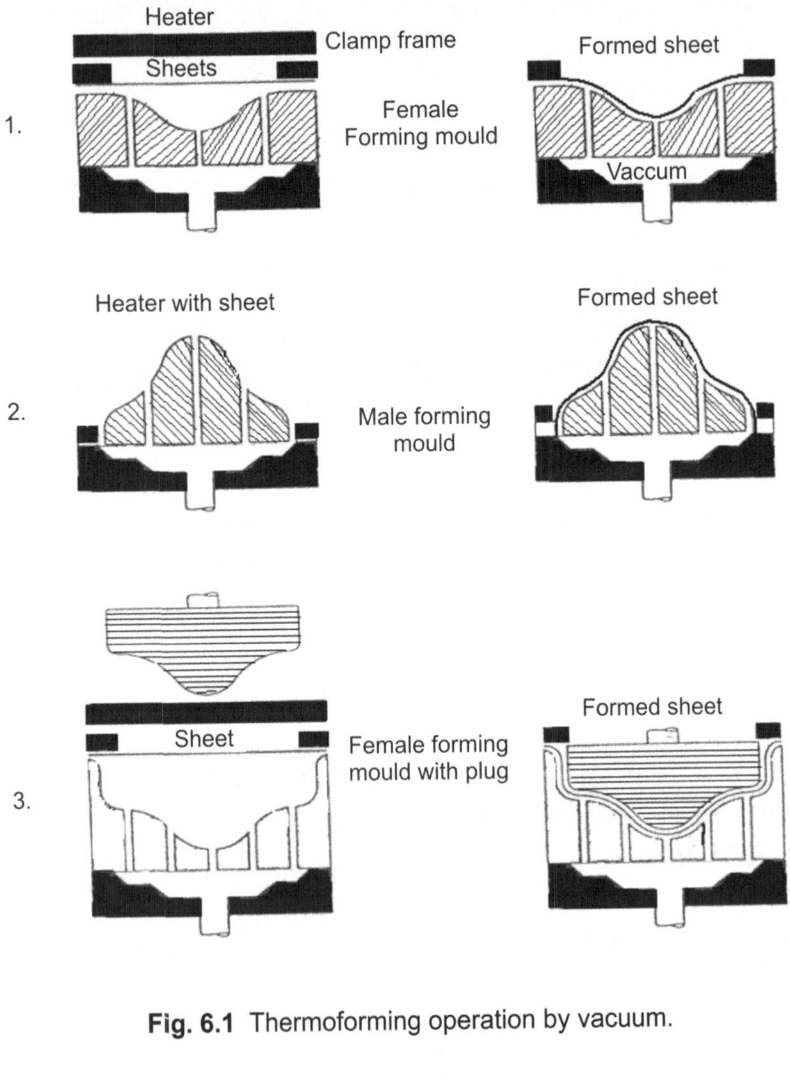

Fig. 6.1 Thermoforming operation by vacuum.

Feeding operation

Feeding mechanism largely depends on the product being fed, i.e. ampoules, tablets, lozenges, soft and hard gelatin capsules, suppositories, drops, etc. size and the number of tracks across the web. Uncoated tablets and capsules are normally fed from a vibratory bowl via channels or tubes by gravity. Vacuum extraction is frequently applied to the bowl, tubes etc. to minimize powder and tablet chip which may finish up in the seal or tray. Conventional flat – shaped uncoated tablets usually present no problems, but sugar coated or beveled tablets with little side edge thickness may cause difficulties due to dovetailing or shingling. In such cases a rotary table feed may be used.

Filling operation

Generally a speed of 250 items per minute per track is considered good. A flood and sweep fill may be employed for sugar – coated tablets when speeds in excess of 300 per track are possible. These speeds are interrelated to the physical restrictions of feeding an item by gravity. Reciprocating tube fillers and chevron roller fillers are used on some machines where the close proximity of the tracks would cause restrictions to the filling speeds.

Heat sealing

Heat sealing may be based on platens or cylinders or rarely a combination of the two. Irrespective of the process the pocket must be sufficient to allow a clearance between the product and the lidding (normally around 0.5 mm), otherwise the product may stick to the lid on heat sealing. The platen requires two uniformly flat surfaces with the base platen – shaped to accept the blister. On certain machines one platen applies the heat via thermostatically or electronically controlled and the other is cooled by air or water to facilitate rapid and effective sealing with minimum of web distortion. The platen process usually operates at a lower temperature and longer dwell time than cylinders, but can suffer from transferring more heat via the web to the product. On other platen machines the web passes thought pre-heaters (Fig. 6.2).

Uniformity of seal depends on the ideal matching of two platens, which under well-controlled conditions certainly gives a better seal. The pattern of seal may be line, pyramid or cross-hatch, but the last of these is always preferred. The seal is achieved at the peripheral contact between two rollers, one of which is heated. As the process stay time is very short, a higher temperature usually has to be employed.

The seal pattern is only on the heated platen or sealing roller, with the cooling platen or cylinder being smooth/flat. The success of either type of seal is inspected by the uniformity of the seal impression, by vacuum tests or material deflection under vacuum/ pressure.

The most occurring problem associated with the heat-sealing operations is 'curl' into the tray, thereby creating problems for automatic enveloping cartoning. Theory suggests that the different coefficients of expansion for foil and film are partly the cause, and film shrinkage due to molecular reorientation also support curl. Curl can be reduced or overcome by incorporating thermoformed ribs in the tray; or by reserving the web curvature by passing over a tension roller or through radius guides. With soft foils the problem of curl is less as it stretches more readily. The curl can also be reduced by between-pocket perforations or cuts.

The degree of curl depends somewhat on

- The type of machine employed
- The type of web
- Design of the tray

Punching out or guillotining the tray from the web

Exclusion of the sealed trays from the web by punching out or die cutting enables shaped trays without sharp corners to be produced. Guillotining, if used, always produces sharp corners which if not carefully handled may penetrate pockets of trays. Punching –out (unless a reciprocating or rotary unit is employed) is usually carried out as an intermittent action with platens. In the case of continuous motion machines the web is usually converted to intermittent by a 'slack loop' prior to either punching out or guillotining. However, continuous motion can be retained by rotary die cutting, which is certainly more costly in terms of dies and equipment. Wastage can be minimized in platen punching by staggering the action i.e. 2 then 1 in a 3 tray across the web configuration. Reciprocating cutting forms can also maintain a continuous motion hence increase output speeds.

Product check/rejection (Quality control)

Prior lidding inspection offers mechanical, electrical sensing or an imaging technique, while after lidding offers either infrared or X-ray methods. Product inspection prior to lidding is performed for problems when the web jerked or slowed down/accelerated so that items could be thrown out of the pockets. Although this aspect has now been eliminated from most machines, the after-lidding option had further problems when clear material were replaced with dark tinted or opaque or foil/foil blisters to improve either child-resistance or moistures, the light, oxygen protection. Thus depending on the machine around web materials, the original before and after lidding options remain.

Between-pocket perforation or similar options present a means by which packets can be readily separated and / or the prevention of any opening method from exposing more than one product. In the US-type peelable packs, perforation or something similar is usually essential to control the peel. Perforation may be achieved by true perforation (a series of bridges and cut-through zones); a part cut thought by sharp blades or knives, or the use of a form of laser technology. Such operations usually require a larger between pocket distance (3 to 4 mm increases to 5 to 7mm), adding to the overall size of the blister. The operation s normally carried out either prior to tie punching out or within the punching, guillotines stage. It must b performed on well-cooled trays, as warm or flexible materials are less easy to penetrate and can blunt the knives more easily.

While evaluating protectiveness of blister packs, detailed knowledge of the product, the packaging material and the process to be employed is an essential requirement. Blister packs are unlikely to give sufficient protection to product severely affected by moisture, oxygen, carbon dioxide, as all known plastic material are to some degree permeable to these factors. A full pack evaluation needs to consider a number of requirements like:

1. Atmospheric factors e.g. oxygen, moisture, carbon dioxide, microbiological hazards, which may result in product deterioration,
2. Product ingredients migration from within the product to the outside atmosphere
3. From the pack into the product migration or reaction between the product and pack, e.g. discoloration, softening of the heat-seal.
4. Damage to pack or product due to mechanical stress
5. Shelf life protection requirement of the pack.

Factor affecting in Machine choice

The choice of blister machine is based on number of factors. These factors include:

1. Variety of material, used for lid and tray
2. Variety of feed, number of feed tracks
3. Available area per stroke on machine (web and length)
4. Trays size and configuration
5. Nature of product
6. Specific nature of blister wall distribution and seal effectiveness
7. Blister type (push – through, peelable)

8. Machine specifications e.g. vacuum, pressure, mechanical plugs assistance etc.
9. Type of heating available – radiant, conducted, preheat, etc.
10. Type of removal from web, punch or guillotine etc.
11. Need for perforations, scores or slits between pockets.

Based on the products and associated problems, the machines may be designed to specific functions. Most common machines generally offer fixed platen or cylinder sizes. These can be incorporated with printing of the lidding by gravure or flexography. Batch marking can be achieved by printing, embossing or de-bossing. Various types of product detection systems are available.

Design for tray and blister

Proper designing of tray may avoid the wastage and may lead to maximum output of blister. Designing trays is an important feature and it explains number of theoretical aspects to be taken into consideration. These are discussed below:

- Square or near right–angled corner or bases should be avoid in blisters as they bring about difficulty in release from the mould and lead to thinning in those areas. Equally, undercuts should also be avoided.
- Vacuum forming is associated with less uniform wall distribution and so this can be overcame by changing the angle of the draw from 6^0 to $12^0 – 20^0$, but this may add to cost if the tray size is increased to a point where output/ tray per area is reduced. Similarly, pressure plus plug assisted forming is usual where a draw angle of 3^0 - 6^0 is taken, as the draw angle is less significant with pressure.
- Satisfactory seals can be obtained with a separation of 2-3 mm space between blisters but there may be some risk of exposing the neighboring pocket when a product is pushed out. Therefore 3-4 mm is more common for both between blisters and margin seals.
- With in-between pocket perforations, this distance becomes 5-7 mm.
- For peelable blisters it is essential to enlarge the peelable edge by at least 5mm. alternatively, a cut-away part of the blister tray either internal to the design or on a seal can be employed to facilitate a excellent for the peel feature.
- In case of tray of the push-through type without perforations providing minimum wastage and maximum machine output, any

modification to a larger tray (perforations, or perforations plus peel, etc.) will lead to fewer trays per stroke, more wastage and lower output.

- Blister trays with properly designed perforations will meet up the US protocol for child resistant non-reclosable packs. Children tended to be satisfied by tearing along the perforations rather than penetrating into the pack. Dark tinted or opaque plastics are essential to render the item contained therein less attractive.

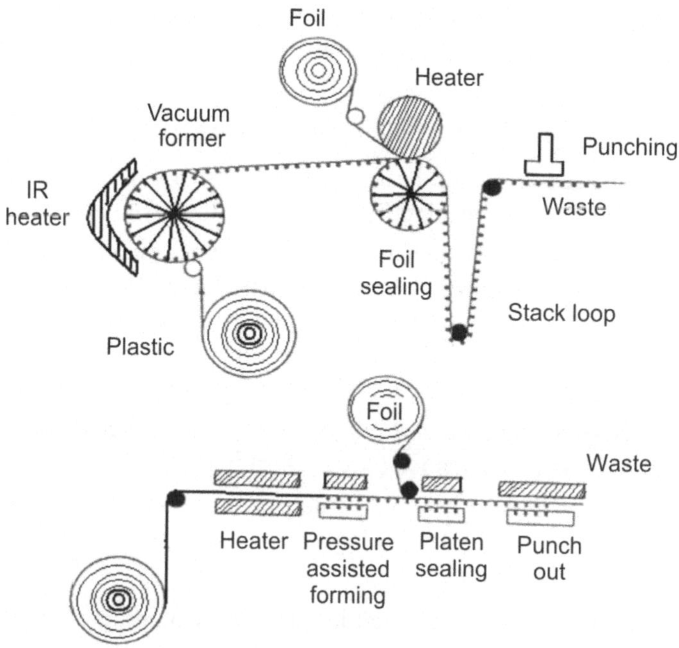

Fig. 6.2 Blister machines.

Cold forming

In the case of cold forming, an aluminum-based laminate film is simply pressed into a mold by means of a stamp. The aluminum will be elongated and maintain the formed shape. In the industry these blisters are called 'cold form foil (CFF) blisters'. The principal advantage of cold form foil blisters is that the use of aluminum is offering a near complete barrier for water and oxygen, allowing an extended product expiry date. The principal disadvantages of cold form foil blisters are: the slower speed of production compared to thermoforming; the lack of transparency of the package (a therapy compliance disadvantage); and the larger size of the blister card (aluminum can not be formed with near 90 degree angles).

Laminations of plastic and foil, commonly of 40μm along with foil, can be physically formed on modified blister packing equipment by a cold mechanical forming operation, which is carried out between male and female dies. These laminations of plastic and foil allow foil to flow without flex cracks during forming although limitations in draw angles and forming depth still remained. This problem can be reduced (but not totally overcome) by using a modified Aluminium alloy, the selection of the laminations, the forming design and the best machine factors. The forming material now consists of such combinations as a biaxial oriented plastic laminated to 40 to 60μm, Aluminium foil and an internal (third) layer of plastic (PE, PVC). Cold forming is also being achieved with multilayer material containing two layers of soft foil separated by a polymer layer.

The basic thoughts behind Cold forming process are:

1. Selecting a foil, which has been embossed or finely creased; can be extended by air or mechanical pressure without showing flex cracks.
2. Selecting a reel of material with regular cross- direction slits (as used on servac suppository machine). A male/female mechanical forming operation is carried out between each slit. This mechanical operation forms the foil and the slit area moves (opens), thereby preventing any high degree of stress (uses 12.5μm OPP/40μm foil / HSL)
3. Adhering the material and carrying out a true punch action where the non- held area is extended (stretch forming)
4. Double forming operation, which reduces the tray size to 20% (Advanced forming Technology (AFT) process). These tend to use Teflon stretching dies.

The Foil originally is a metallic material which have a crystalline structure and commercially pure grades of aluminum foil had a grain size of around 40μm and so it becomes difficult to make 40μm and less gauges suitable for cold forming. The foil of 40μm was likely to be a mono- grain layer. By the production of special foils of smaller grain size, new 40 to 60μm foils consisting of multiple layers can more readily stretch without any fracture risk. Laminating this foil to biaxially oriented polypropylenes, polyamides and polyester film further added to the stretch forming potential.

It is reported that the better the laminating bond between the foil and the plastic, the better will be the forming operation. To date a typical forming ration is approximately 1:2.5 (i.e. a cavity diameter of 20mm would have a depth of 8mm), using polyamide / aluminum 40μm / polyamide / PVC film.

Two selected techniques have been engaged in the development of the technological progress:

1. MAD unit developed by Bosch (a machine to measure multi-axial elongation)
2. Hoogoven screen – a specially printed screen from which stresses and deforming during forming can be quantified (also useful for plastic films)

The factors, which control the stretch forming operation, comprise.

1. Gauge and type of foil used
2. Fineness of the crystalline structure of the foil
3. Design of the formed area including angle of draw and the depth of draw (width to depth currently does not exceed a ratio of 2:5)
4. Structure of the supporting or carrier lamination including the uniformity of the oriented plastic.
5. Positive lamination between foil and outer ply
6. Forming speed and the friction properties of the laminate
7. Quality of the die, especially the surface finish.

Tropicalised lidding

Tropicalised lidding may be achieved by a performing operation, thus allowing a seal with somewhat narrow margins or by direct lidding (no stretch) with a more gradual 'draw' and wider margins. It is an alternative ways of obtaining excellent moisture protection over relatively expensive cold forming. Putting a foil lid over the tray section and sealing it to the blister margins, known as the 'tropicalised' blister.

6.3 Strips Packaging

The strip packaging may be used to package any element which is relatively infrangible. Thus, while the packaging may be particularly appropriate for pharmaceutical products, it can also be used for other products where safety aspect is a factor.

These packs consist of one or two plies, made from regenerated cellulose, paper, plastic, foil or a combination of these. The dosage form is inserted into a pocket area against a recess in a heated platen (or roller) or may be inserted in preformed pockets. This from of packaging is used for packaging of tablets and capsules. Two sheets of film are used for making a strip package and the sheets are fed through heated crimping roller. Λ continuous strip of pockets are formed and cut as required. The product is sealed between the two sheets of film along with a seal around each tablet/ capsule and perforations for easy cutting.

The use of high barrier material like foil, or saran – coated film with excellent seal formation, enables this packaging suitable for moisture – sensitive products. Other lamination found suitable for strip packaging is a four fold one consisting of paper/ polyethylene/ foil/ polyethylene. Usually a cellophane film, on accounts of its heat sealable nature and transparency is also used, especially when the visibility of the product is important.

Strip packs present an alternative form of pack for a unit dosage. Strip can be produced from single or multi – ply material provided the two inner plies could be sealed by heat or pressure (e.g. cold 'self adhesive' seal). Material can range from relatively permeable plies to those, which incorporate a foil ply of sufficient thickness (and effectiveness of seal) that an individual hermetic seal is produced for each dosage. To date strip packs are usually produced at lower speeds and also occupy greater volume than used, output speed and item size.

Strip packaging process

As compared to blister packaging units, strip packaging machines are much simpler and smaller, consisting of a feed system, product insertion plus heat sealing and a guillotining operation size. Mostly a strip pack can be created by introducing an article, which extends a pockets area during insertion, or by a performing filling operation as illustrated in Fig. 6.3 and 6.4. The latter method gives less strain (or more controlled forming) to the pocket area and reduces the material needed by 20-35%. By this process either one or both sides of the plies may be mechanically formed, but it can only be applied for materials which will 'stretch' without tearing.

Material Feeding is usually through a vibratory bowl with feeding tracks (usually up to a maximum of sixteen). A rotating table plus drop or sweep may also be used. The majority of machines employ a vertical feed (dropping by gravity) but seldom the web is run horizontally with a platen type sweep. The pocket area is created recesses either in a platen or more usually in a heat-sealing cylinder, where a circumferential point seal is made between two inter meshing cylinders.

Similar to blister packs, the maximum speed depends on the size of the article and gravity. A maximum speed of 250 – 300 per track is likely with a 325 mg (five grain) type of aspirin product. Removal of powder, chips, etc. is achieved by vacuum removal.

Either a scissors or guillotine motion or rotary die cutting invariably does cutting of the emerging web.

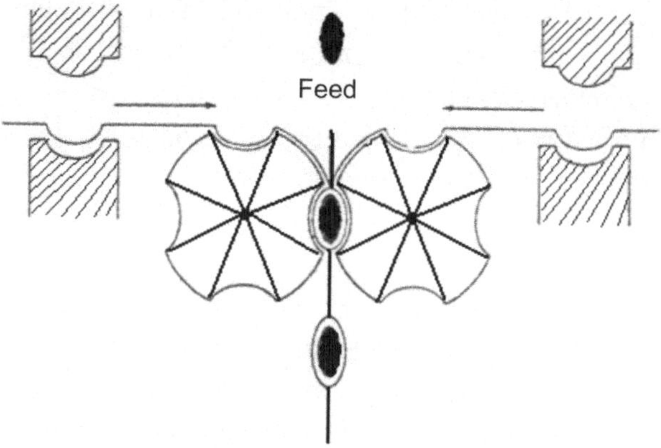

Fig. 6.3 Strip Pack with preformed packet.

Further stages include printing, perforating, batch coding, magic eye registration, etc. It is destined from blisters, perforation does not usually add to the seal width, as pocket seals are normally 5 mm or more. Two separate webs are used by most machines but occasionally a single centrally folded web may be employed. Strip packaging is closely related to sachet packaging and in certain cases it is difficult to distinguish between the two. Two different plies can also be used (top and bottom) provided the sealants are compatible.

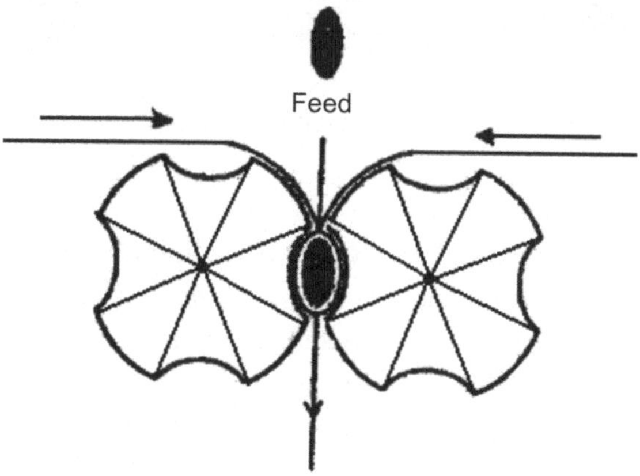

Fig. 6.4 Strip pack in Item formed Pockets.

Machine capacity

Strip packaging machines with maximum output around 2,000 items per minute (eight–tracks machines) are common. More ever, four tracks, two tracks and single track are more usual with outputs of 1,000, 500 and 250 items per minute repetitively. In recent times sixteen and thirty-two track machines have achieved output of 4,000 and 8,000 items per minute. These are also have an integral cartoning option.

Strip designing

Generally strips are available in very basic designs, as the emerging units are constantly rectangular or square strips. However, the pocket portion can be round, oval or square. The shape, size and area of pocket are dependent on the shape, thickness and diameter of the product (Fig. 6.5). In case, the pocket is too small or 'tight' tearing damage of the pocket edge or wrinkling of the seal region may occur. The minimal seal width may be kept as low as 4 mm, but frequently 5mm and above is employed. If the seal region is likely to wrinkle or crease then wider seals may be necessary.

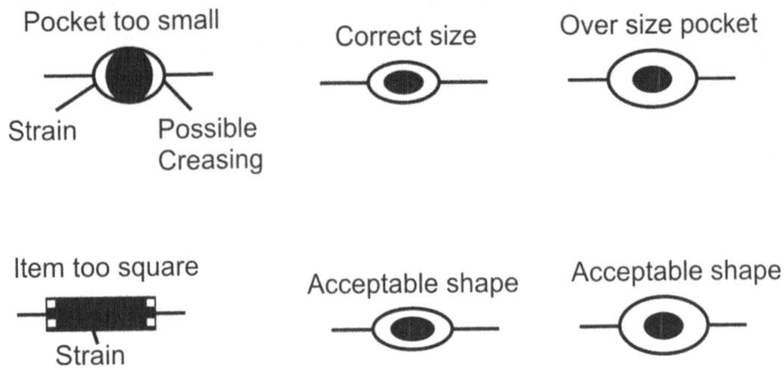

Fig. 6.5 Strip pocket designs illustrating acceptability of shape.

The cylindrical–type-sealing process does not generally have a separate cooling cycle, hence any pull on the seal ply will be inclined to weaken the seal as the sealant is still pliable. It becomes important with foil that where a small seal margin is used, it may be necessary to have air-cooling cylinders. Further we know that, the warm materials are commonly more difficult to cut than cold materials. This may bring in a need for extra cooling device or moving away the cutting operation from the heat-sealing area; as a 'soft seal' may avoid a clean cut. Alternatively, the use of cold seals (Self adhesive materials) is increasing mainly due to such materials can be sealed at higher speeds than usual heat seals.

6.4 Leakage Testing and Package Integrity

It is mandatory to both blister and strip packs to carry out the check for pack integrity. It is of vital importance, both to guarantee product shelf life and to ensure that the product has not become 'contaminated' from some environmental sources. While speaking about contamination, it referred to as the deterioration of quality of product inside pack in its organoleptic, physical, chemical or biological nature.

Desirable pack integrity can be achieved by making control over machine operations or in-process quality control on the packs produced. Mainly two principles are being involved for checking package integrity:

1. ***Destructive type testing:*** this type of testing equipment is based on destruction of seal by applying vacuum. These tests include vacuum tests under water and burst tests. Vacuum tests under water gives the following seal integrity test:

   Dip the packages under test in a pot containing coloured water (15-25^{0}C) and place the pot in the vacuum chamber. Apply the appropriate vacuum of 33 kPa (250mm of mercury) for strip packages or 24kPa (180mm of mercury) for blister packages, for 30s. Return to atmospheric pressure and remove the pot from the vacuum chamber. Take away the test package from the pot and blot off the excess water. Examine the package for ingress of water into the pockets.

2. ***Non-Destructive type testing:*** This type of testing equipment is usually based on a dry pressure vacuum procedure followed by detection of pack distortion (deflection) or non-distortion (non-deflection), i.e. packs with effective seals become concave them convex as positive pressure changes to negative pressure, while leaking packs either do not change or show less or limited distortion, depending on the scale of the leakage. The introduction of non-destructive tests should therefore not only improve output, but enable better feedback on machine performance. This also means that where seal integrity is lost or is suspect more effective corrective actions can be undertaken, including improvements in on- line controls.

Pinholes and Package integrity

Pinholes is the common feature of aluminum foil. Pinholes normally refer to the minute holes, which may be present in the foil. Usually the foil of

0.017 mm caliper and above is recognized as commercially pinhole free. The foil with 0.025 mm can normally be 'guaranteed', Pinhole free. Foil below 0.017 mm gradually shows an increasing number of pinholes. However, it has been noted that the water vapor permeation (WVP) of foil of 0.006mm is lower than any nominal gauge of plastic when used in a heat sealing lamination. When foil is laminated or coated the initial pinholes tends to be 'filled in' thus reduces any permeation risk till further. Permeation is then related to the plastic, which covers the pinholes.

The blister offers extremely low permeability via the side of foil, hence permeability relates mainly to the plastic of tray and the wall distribution. In case of strip packs permeation may occur through edge seals, or more likely capillaries created by creases in the edge seals, as well as actual foil perforations in general pinholes and perforations cannot be detected by vacuum dye tests, as the plastic plies normally remain intact. Capillary leakage can be detected by this method, it may be assisted by adding a wetting agent to the solution.

6.5 Newer Approaches in Blister and Strip Packaging

The most important issues in blister and strip packaging are creating the child–resistance and moisture-oxygen protection packs. Recently these goals are being achieved by improvisation in material, machinery and their performance, although each of these may be influenced by the properties of the product, costs, and change–over time, output, etc. Some of the recent innovations are as follows:

(i) Innovations in Machine improvements

The majority of machines, which are now based on pressure forming with plug assistance, can be adapted to cold forming. A machine has a tendency to occupy less length by working over two levels or by doubling back on itself. Though for conventional blisters, maximum speeds of over 6,000 items/minute are still relatively rare, much higher speeds are predicted. Some of the machine improvements features considered recently include:

- use of larger reels,
- various forms of preheat with variable temperature ranges either side of web,
- preference for pressure forming with plug assistance,

- improve cooling of moulds,
- possible preheat for foil sealing section,
- automatic splice facility.
- removal of the blisters from the web with the minimum of wastage
- modular machines and Servo technology driven systems are also being offered.
- de-blistering units (on and off the machine) are more readily available
- software controlled blister packaging equipments

Both semi-automated process and fully automated process machines are available for effective blister and strip packaging. Machines are equipped with deferential features to give a protective over wrap against physical damage, to prevent ingress or egress, to aid display or to act as a tamper-evident and tamper-resistant feature.

Semi – automatic machines also imply the same principles, sheets may be used instead of reels. The types of material and the process in use tend to be based on their cost-effective and strength characteristic. Where a tray or container is not necessary for a direct form fill seal operation, the thermoformed item can be made-up either in a female mould or over a male mould. The different combinations of vacuum, pressure and mechanical (plug assist) are indicated in drawings. In case of vacuum forming and where a female die is used the process is referred as "straight vacuum forming". Where a male die is used, the process is termed as "drape forming". Greater pressures may be employed in case of air pressure forming.

The majority of the individual type of blisters is made by vacuum forming either into or over (Drape type) moulds, which are frequently made from metals such as aluminum. The selection between female and male moulds depends on a number of features, as discussed below in Fig. 6.6.

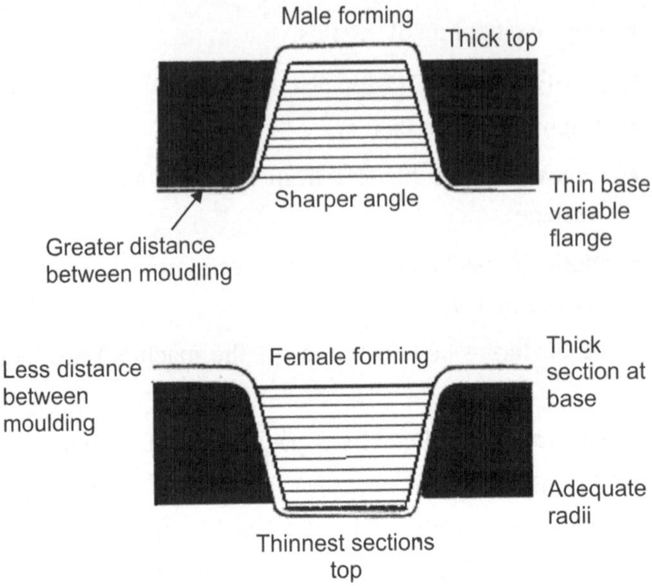

Fig. 6.6 Vacuum forming: Male and female type.

Female mould features

- Base (flange) is thickest, and uniform;
- Top of mould is thinner.
- Forming can usually be put closer together hence more moldings per sheet.
- No webbing between moulds,
- Limitations in depth of draw,
- Good radiused corners required–not suitable for sharp corners.

Male mould features

- Base (flange) is thin and non- uniform;
- Top of mould is thickest,
- Capable of greater depth of draw for a given caliper
- May tend to web if layout is not properly designed.
- Needs greater space between moldings (rule of thumb 1.5 X draw height), therefore fewer moldings per sheet are achieved.
- It is capable of achieving relatively sharp corners.
- Plug assistance can be used, particularly where webbing is likely to occur, but this generally slows output.

- Air holes for venting vary between 0.50 mm and 1mm for most small moldings.
- Moldings can be 'ribbed' to improve strength or to enable a thinner material to be used e.g. 300μm reduced to 200μm by ribbing. Improve wall distribution can also be achieved by extending the softened plastic sheet by vacuum or pressure prior to final forming (often known as balloon or bubble extension).

A few comparatively easy blister packaging equipment is produced on a rotary table principle whereby a blister can be formed, hand filled, lidded and the completed blister pack removed after 90^0 movements. Few more work on a reciprocating bed principle of forming and lidding. Above methods are often used for optic lenses and ophthalmic-type solution for lens cleaning.

A number of typical blister packs are shown in Fig. 6.7 below utilizing various assembly and sealing methods e.g.

- Blister locked between cards by adhesive or staples
- Folded card- plus heat seal, adhesive or staples.
- Blister locked to card by label.
- Heat seal through lid (board, laminate or label)
- Heat seal through plastic
- Heat seal through plastic and lid

Blisters machine for liquid and semi liquids

Blister machines further offers improvisation for packing liquid and semi-liquid Pharmaceutical preparations. The conventional horizontal formed blister can be used for filling semi-viscous to viscous products or sometimes for smaller quantities of liquid, but it is essential to turn them to a vertical position when a higher fill volume per pack is to be achieved. The filling single or double-sided formed blisters in the vertical position can be performed in cut individuals, cut sticks or an occasionally or constantly reel-fed system. Preliminary attempts to employ a double-sided formed blister led to low operational speeds. The beginning of laminate or coated webs whereby the sides of each web can be heat-sealed led to significant increases in output. Today machines are available which can produce packs with liquid volume fills between 1 and 90 ml with output speeds of up to 20,000 an hour. Some machine manufacturers also offer machines, a development of their suppository packs, where preformed blisters are delivered in reels.

In theory machines of these types should be capable of aseptic production and hence in the long term could compete with other machines, which operate on aseptic systems.

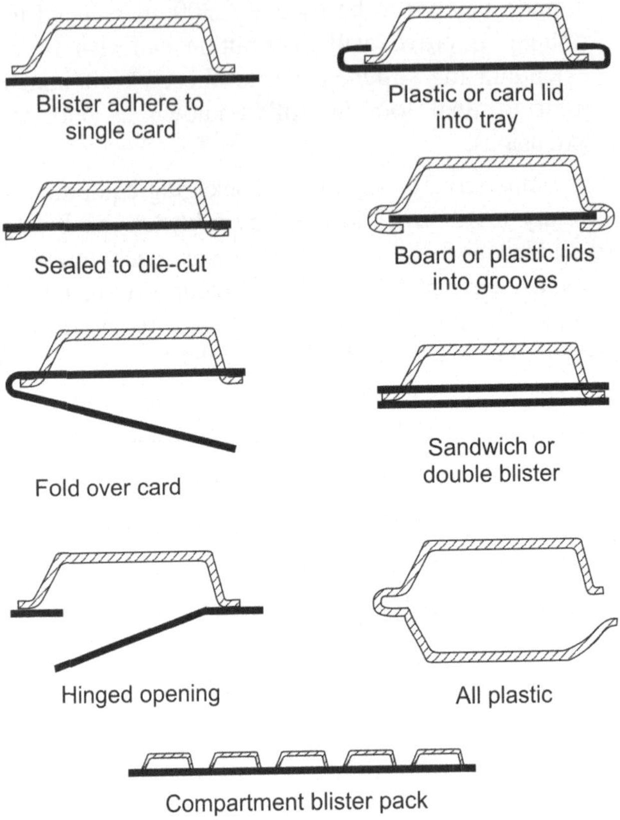

Fig. 6.7 Thermoformed Card Pack.

Tamper-resistance and child-resistance

Since the introduction of child-resistant packaging, in 1970s in the UK, a noteworthy decrease in the reported cases of child poisoning instances has been observed. These resistance packs has involved those re-closable CRCs, which have accepted a test using, children and non- re-closable packs, which have not passed any test but have exposed a remarkable growth in use. Although a variety of packs are in use where strips, sachets etc. are inherently child safe, there is still a need to improve the more 'flimsy' forms of pack. In contrarily many child-resistant pack still present opening problems to the ever-increasing elderly population who are on regular medication in order to stay alive. The overcoming opening problem may exposing both the product and children at risk. The vital issue is still the challenge to industries.

(ii) *Innovations in Materials*

Most of the cold-formed materials started with 40µm foil and utilized a relatively shallow well-ready used draw. There occurs perforation problems with such foils. Hence 45-50µm foils have been substituted with gauges available up to 60µm.

More recently Aclar copolymer based material using 22A, 33C, 88A have been challenged by the homo-polymer R_x series with lower costs joined to good moisture barrier properties. However, there are many alternatives over wrapping systems, which may be used to improve moisture barrier. The Japanese have just introduce silicon oxide coated PET over wrap. For desired clarity PVC, PS, PET, or cellulose acetate may be used. The last of these tends to cost more even though the molding cycle is approximately 25% faster, this does not compensate for the higher material cost (+30%). PET is often preferred for surgical or medical device systems.

(iii) *Innovations in Extreme barrier packaging*

In extreme barrier packaging, this traditionally means either the 50 µm (2 mil) Aclar structures originally developed for the U.S. market, or three-ply CFF structures originally developed in the European market. However, users are demanding ever more extreme barriers, especially against moisture. One can only understand extreme barrier packaging by understanding materials, machinery and tooling. Three distinct trends drive this demand for more barrier, and will become even more important in the future:

- The growing importance of drug delivery systems;
- The tendency to develop active pharmaceutical ingredients (APIs) that are highly sensitive to moisture and oxygen;
- Use of "super disintegrants" such as sodium starch glycolate, which are becoming more important in formulation. These materials are increasingly being used in generic drug manufacturing, to help achieve desired dissolution curves and enhance bioavailability. Unlike traditional excipients, these materials require packaging materials with a much higher moisture barrier.

Until now, this demand for packaging with improved barrier properties has been met mainly by enhancing existing products, or combining materials. Each of the four traditional forming material technologies (CFF, PCTFE, COC, and

PVdC) is available in new forms that address the growing demand for extreme barrier properties.

- ***CFF/Alu Alu:*** New four-and five-layer structures reduce any risk of delamination and pin-hole, and enhance stiffness and processing speed.
- ***ACLAR/PCTFE:*** Films are now available up to 100 μm or 4 mils thick, doubling the barrier offered in the traditional 50 μm or 2-mil structures.
- ***TOPAS/COC:*** New combinations of COC with PVdC are pushing barriers obtainable with these structures into barrier ranges formerly available only from CFF/Alu Alu or Aclar/PCTFE
- ***PVdC:*** Solvay is working on a new product with several of its converters that it says significantly enhances the barrier obtained from a given gram coating weight.

CHAPTER 7

Quality Control in Packaging

Summary

7.1 Introduction

Quality is the backbone of any pharmaceutical industry, regardless of whether it concerns with the quality of medicines or packaging material. Compromise in the quality assurance methods can lead to serious implication. The defects in packaging may have harmful affect on the dosage form. The only way to ascertain total quality in pharmaceuticals is to make quality evaluation in all respects of active ingredients and additives and that of packaging materials in which they are packed.

7.2 Basic Requirements for Commonly used Dosage Forms

The package must also contain relevant information regarding the product, storage condition, administration and use, warning along with other information required by the regulatory authorities. The protective envelopes should be sufficient robust to withstand normal handling. On removal its protective envelope the container should show no signs of leakage and no growth of microorganisms.

7.3 Tests on Containers

USP has prescribed various tests on the units as well as multiple dose containers such as hydrolytic resistance test, permeation test, light transmission test etc.

- **Testing of glass containers**

  Generally hydrolytic resistance and chemical resistance tests are carried out on glass containers. According to USP XXVII, powdered glass test, water attack test and test for arsenic are carried out on glass containers.

1. Powdered glass test
2. Water attack at 121^0c

- **Testing of plastic containers**

  Indian Pharmacopoeia prescribes Leakage test and Collapsibility test for plastic containers. These tests are applicable for both injectable and non-injectable. Water vapour permeability test of Indian Pharmacopoeia is applicable for injectable preparations. Other tests for plastic containers include physicochemical tests, light transmission test, compatibility evaluation, and safety evaluation.

  1. Leakage test
  2. Collapsibility test
  3. Water vapour permeability test
  4. Light transmission test
  5. Compatibility evaluation
  6. Safety evaluation

Physicochemical tests: *These includes*

- *Nonvolatile residue (NVR):* NVR measures organic / inorganic residue soluble in extraction media [limit – not more than (NMT) 15 mg.]
- *Residue on ignition:* This test is performed when the NVR is greater than 5 mg.
- *Heavy metals:* This detects the presence of metals such as lead, tin, zinc, etc.
- *Buffering capacity:* This measures the alkalinity / acidity of the extract.

7.1 Introduction

Quality is the backbone of any pharmaceutical industry, regardless of whether it concerns with the quality of medicines or packaging material. Utmost quality assurance comprises central theme of the total approach to Good Manufacturing Practices. To ensure high quality of pharmaceutical packs, the quality management system must take into consideration the necessities of the local authorities and the legislation, the raw material, the product, the production process, and the policy of the manufacturer. Compromise in the quality assurance methods can lead to serious implication. The defects in packaging may have harmful affect on the dosage form. The only way to ascertain total quality in pharmaceuticals is to make quality evaluation in all respects of active ingredients and additives and that of packaging materials in which they are packed.

Designing

The foremost requirement for packaging must be a good design as relevant to the need of the product, product-pack stability, compatibility, manufacturing and distribution system, the patient, and the legislation.

Product / component stability and compatibility parameters to be studied include moisture and gas protection, light and temperature protection, microbiological integrity, pH stability, and the migration of components from the medicines to the packaging components including preservatives, volatile actives, stabilizers, plasticizers, anti-oxidants, etc. The QA team needs to stress on the physical aspect of production and marketing e.g. component dimensions (compatibility with machines), machinery (line speed, etc), stock policy (batch sizes, etc.), distribution system (domestic or export, weight, friability, etc.) legislation (fill weight, labelling, etc.), aesthetic aspect and supplier (limitation, requirements, etc.).

Quality control

The quality control (QC) test specification must be built around the component specifications and should be designed to achieve consistency through-identifying the product, market and production requirements, defining and rating the acceptable quality levels, ensuring sampling, and making a realistic test regime.

The QC testing programme depends upon the company policy, type of material/container, the market use, and the supplier confidence and production requirements. The laboratory routine analysis should include identification, visual inspection, quantity, dimension test, physical test, chemical tests, microbiological test, component and performance measurements, defect classification and action plans and records.

Sampling

Sampling is used to check the correctness of the label, packaging material, container reference, as well as in the acceptance of consignments, detecting adulteration of the medicinal product, obtaining a sample for retention, etc. Effective sampling must take into account the homogeneity and uniformity of the material so as to ensure that the sample is representative of the entire batch.

Inspection and audit of suppliers

The primary goal of inspection is to ascertain the quality of the packaging. These include, documentation, storage of starting material and finished product, validation of programmes, production and in-process controls, calibration of instruments and measurement systems, control of labels, etc. National/international authorities audit pharmaceutical manufacturers. All suppliers of pharmaceuticals and packaging materials play an important role in the chain of quality assurance of the final medicinal product.

7.2 Basic Requirements for Commonly used Dosage Forms

The package must also contain relevant information regarding the product, storage condition, administration and use, warning along with other information required by the regulatory authorities. General labelling information for a product is shown in Table 7.1.

Table 7.1 Labelling requirement of a drug product

- Name of product
- Quantity
- Content
- Batch no
- Mfg. date
- Exp. date
- Directions for use
- Storage conditions
- Special instructions
- Name and address of manufacturer

Tablets

The following are the labelling requirements for tablets:

- Store in well closed containers protected from light, moisture, crushing and mechanical shock,
- Special tablets e.g. effervescent tablets should stored in tightly closed containers/ moisture-proof packs, and
- Effervescent tablets should be labeled, "Not to be swallowed"

Capsules

Capsules must comply with the labelling requirements as under:

- Store/ pack in a manner that protects them from microbial contamination,
- Keep in well closed containers,
- Protect from light, excessive moisture, or dryness, and
- Do not store above 30 °C

Parenteral preparations

Unless specified otherwise these should be stored in containers with sufficient transparency to allow visual inspection:

- Container should not adversely affect the quality of the product by either leaching or extracting product components,
- Closures should be inert,
- Closures for parenterals should have firm seals to avoid contamination/entry of microbes,
- Elastomer of the closures should allow piercing of needle through it with least shedding, and for multi-dose containers, the closures be sufficiently elastic to allow reseal after the needle is withdrawn.

Topical semi-solid forms

- Containers for these preparations should be made of material which does not adversely affect the quality of the preparation,
- Container material does not allow diffusion of any kind into or across the container,
- Closures for such preparations should be designed to minimize microbial contamination,

- Container should protect the preparation from light, moisture, damage during handling and transportation, and
- Closures for such be tamper-evident,
- Preparation for nasal, aural, vaginal or rectal use should be supplied in containers adapted for the appropriate delivery of the product to the site of application, or should be supplied with an applicator.

These preparations should be stored in well-closed containers. The preparation should maintain its integrity throughout the shelf-life and should be stored at temperature not exceeding 25 °C.

Aerosols

USP specifies certain warming(s), which should appear on the aerosol containers:

- Avoid inhaling pressure. Do not puncture of incinerate container. Do not expose to heat. Store at temperature not above 12 °C. Keep out of reach mucus membrane,
- Do not inhale directly; deliberate inhalation of contents can cause death.
- Use only as directed; intentional misuse by deliberately concentrating and inhaling the contents can be harmful or fatal.

Packaging of blood and related products

Special consideration has to be given for packaging of blood and its related products. Some of the salient points mentioned in the USP XXVII are listed below:

1. Plastic containers for the collection, storage, processing and administration of blood and its components should be sterile.
2. In normal conditions of use, the materials of the different parts of the containers should not release monomers or other substances in amounts likely to be harmful and should not lead to any abnormal modifications of the blood. The containers may contain anticoagulant solutions depending on their intended use.

3. The container may be in the form of a single unit or the collecting container may be connected by one or more tubes to one or more secondary containers to allow separation of the blood components to be effected within a closed system.
4. The outlets are of a shape and size allowing for adequate connection of the container with the blood transfusion equipment.
5. The protective coverings on the blood-taking needle and on the appendages should be designed to ensure that sterility is maintained. They are easily removable but are tamper-evident.
6. The containers should be fined with a suitable device for suspending or fixing, which does not hinder the collection, storage, processing or administration of blood.
7. The containers should be shaped in such a manner that when filled they may be centrifuged.
8. The container should be sufficiently transparent to allow adequate visual examination of its contents before and after taking the blood. It should also be sufficiently flexible to offer minimal resistance during filling and emptying under normal conditions of use.
9. The container should contain not more than 5 ml of air.
10. It should be resistant to centrifugation and stretch.
11. It should not show any leakage.
12. Vapor permeability should be very less.
13. It should empty under pressure in less than 2 minutes.
14. It should be resistant to temperature variations.
15. It should comply with sterility tests and pyrogen tests.
16. Sterile plastic containers for human blood and blood components are packed in protective tamper-evident envelopes.

The protective envelopes should be sufficient robust to withstand normal handling. On removal its protective envelope the container should show no signs of leakage and no growth of microorganisms. The labelling requirements for blood and plasma products are depicted in Table 7.2.

Table 7.2 Labelling requirements for Blood and Plasma products.

Whole Human Blood	Plasma
• ABO and Rh groups	• Correct name of the patient
• Volumes of blood and anticoagulant present	• Identifying batch no.
• Date on which blood was collected	• Information concerning potency or concentration
• Expiry date	• Name and address of manufacturer or authorized distributor.
• Required storage conditions	
• Contents must not be used if there is any visible sign of deterioration.	• Date of manufacture • Expiry date
• Number or code by which the history of the preparation can be traced back to the original added bacteriostatic agent.	• Storage conditions • Nature and concentration of any donor.
	• Names and percentages of anticoagulant and other added substances
	• The quantity of water for injection required to reconstitute the original volume
	• Protein content of the reconstituted liquid.
	• Contents must not be used more than 3h after reconstitution.

7.3 Tests on Containers

USP has prescribed various tests on the units as well as multiple dose containers such as hydrolytic resistance test, permeation test, light transmission test etc.

7.3.1 Testing of glass containers

Generally hydrolytic resistance and chemical resistance tests are carried out on glass containers. According to USP XXVII, powdered glass test, water attack test and test for arsenic are carried out on glass containers. The European Pharmacopoeia (EP) provides tests for glass containers including hydrolytic resistance, arsenic, hydrolytic resistance for powdered glass and hydrolytic resistance of etched glass. While the Japanese Pharmacopoeia prescribes test for visual inspection, soluble iron, light transmission and alkalinity.

Powdered glass test

This test is much more stringent than the water attack test. Step-wise procedure of test and sample preparation and test is depicted in Table 7.3 and Table 7.4. In this test the containers are washed, dried, and crushed, sieved and again crushed through a nest of sieves. The portion retained on sieve (excess of 10g) is transferred to a closed container and stored in desiccators. Iron particles are removed from the sample, which is then washed with acetone and dried. 10 g of this crushed glass is mixed with high purity water in a conical flask and exposed to standard regimen of autoclaving at 121 °C for 30 minutes. The water is then separated in another container; to it are added washings of the specimens made using purified water. The pooled sample is titrated using 0.02 N sulphuric acid using methyl red solutions as indicator. The volume of acid used should be in accordance with the values given in Table 7.3.

Table 7.3 Preparation of specimen of powdered glass test

Rinse 6 or more containers → Dry them → Crush into fragments

↓

Divide 100g of coarsely crushed glass into three equal parts

↓

Place one portion in mortar

↓

Crush further by striking 3 or 4 blows with hammer

↓

Nest the sieves

↓

Empty the mortar into 20# sieve (repeat on remaining portions of glass emptying each time in 20# sieve)

↓

Shake the sieve and then remove the glass from 20 and 40#

↓

Crush and sieve again as above

↓

Transfer the retained portion on 50#, which should weigh excess of 10g

↓

Spread the specimen on glazed paper and remove iron particle with help of a magnet

↓

Wash with six 30ml portions of acetone and decant acetone

↓

Dry the contents for 20min at 140 °C

↓

Transfer to weighing bottle and cool in desiccators

↓

Final specimen to be used in powdered glass test

Table 7.4 Powdered glass test (As per USP)

Transfer 10.0 g of prepared specimen in a 250 ml conical flask digested previously with high purity water in a bath at 90 °C

↓

Add to conical flask containing 50ml of high purity water

↓

Cap all the flasks

↓

Autoclave (continue heating for 10 min)

↓

Close vent cock

↓

Adjust temp to 121°C

↓

Hold the temp (121 °C ± 2 °C for 30 min.)

↓

Reduce the heat and wait for autoclave to cool

↓

Cool the flask in running water

↓

Decant water

↓

Wash the residual powdered glass (4 times with 15ml high purity water

↓

Add the decanted washings to main portion

↓

Add 5 drops of methyl red solution

↓

Titrate immediately with 0.02 N sulphuric acid

↓

Record the volume of 0.02 N Sulphuric acid

↓

Volume doesn't exceed that indicated in table for the type of glass concerned

Water attack at 121 °C

The water attack test at 121 °C (USP) is described in Table 7.6. This test is performed on intact containers. The containers are filled up to 90% of their overflow capacity. The containers are exposed to similar autoclaving conditions except that it is carried out for 60 minutes rather than

30 minutes. The contents of the container are pooled and from the pooled contents 100ml of the solution is titrate as in case of powdered glass test. The volume of acid used should not exceed that indicated in Table 7.5.

Table 7.5 Types of glass and their test limits.

Types of glass	General description of glass[a]	Types of test	Limits size[b], ml	Limits (ml of 0.20N)
I	Highly resistant, borosilicate glass	Powdered glass	All	1.0
II	Treated soda-lime glass	Water attack	100 or less	0.7
			Over 100	0.2
III	Soda-lime glass	Powdered glass	All	8.5
IV	General-purpose soda-lime glass	Powdered glass	All	15.0

a. The description applies to containers of this type of glass usually available.
b. Size indicates the overflow capacity of the container.

Table 7.6 Water attack at 121 °C (USP)

Rinse 3 or more containers twice with high purity water

↓

Fill each container to 90% of its overflow capacity

↓

Cap all the flasks, Autoclave for 60 min.

↓

Empty the contents and pool the contents in 250 ml conical flask to a volume of 100 ml

↓

Add five drops of methyl red solution

↓

Titrate with 0.02 N sulphuric acid while warm (complete titration within 60 min.)

↓

Record the volume consumed (corrected for blank obtained by titrating 100 ml of high purity water at same temperature and with same amount of indicator

↓

Volume should not exceed that indicated in Table 7.5 for the type of glass concerned

7.3.2 Plastic containers

Indian Pharmacopoeia prescribes Leakage test and Collapsibility test for plastic containers. These tests are applicable for both injectable and non-injectable. Water vapour permeability test of Indian Pharmacopoeia is applicable for injectable preparations. Other tests for plastic containers include physicochemical tests, light transmission test, compatibility evaluation, and safety evaluation. Some of the important tests given in IP 1996 are shown in Tables 7.7, 7.8 and 7.9.

Table 7.7 Leakage test for plastic containers (Non-injectable & injectable) (IP 1996)

Fill 10 containers with water and fit the closure

↓

Keep them inverted at RT for 24 hours

↓

No sign of leakage from any container

Table 7.8 Water vapour permeability test for plastic containers for injectable preparation (IP 1996).

Fill 5 containers with nominal volume of water and seal

↓

Weigh each container

↓

Allow to stand for 14 days at RH of 60 ± 5 % at 20 ^{0} to 25 $^{\circ}$C

↓

Reweigh the container

↓

Loss of weight in each container should not be more then 0.2%

Table 7.9 Collapsibility test for plastic containers (Non-injectables & injectables) (IP 1996)

This test is applicable for those containers, which have to be squeezed for the withdrawal of the product.

↓

A container by squeezing yields at least 90% of its nominal content at required flow rate at ambient temperature.

Physicochemical tests

The USP physicochemical tests are designed to determine physical and chemical properties of plastics and their extracts, are based on the extraction of plastic material, and it is essential that the designated amount of plastic be used. USP XXVII specifies surface area that must be available for extraction at the designated temperature. The extracting medium, unless specified otherwise, is purified water and is maintained at 70 °C during extraction. The current minimum surface area is 540 cm^2. After the extraction process has taken place, the following checks are performed.

- Nonvolatile residue (NVR): NVR measures organic / inorganic residue soluble in extraction media [limit – not more than (NMT) 15 mg.]
- Residue on ignition: This test is performed when the NVR is greater than 5 mg.
- Heavy metals: This detects the presence of metals such as lead, tin, zinc, etc.
- Buffering capacity: This measures the alkalinity / acidity of the extract.

Primary container / closure system may interact with the dosage form. Extractable and bleachable matter may be generated and may contaminate the product. An extractable is a chemical species that can be released from a container or closure material of construction that has the potential for containing the dosage form. Under certain stress conditions this may be generated through an interaction with the closure system. Where a leachable is a chemical species that has migrated from packaging component or any other component of the product into the dosage form under normal conditions of use or under stability studies. The chemicals that can migrate from the container / closure system include stabilizer, contaminant, lubricant, antioxidant, plasticizer and monomer. The US FDA has been and continues to be the driving force behind the safety evaluation of materials and container closure system. Establishing the safety of container system is of great importance to the medical and pharmaceutical industry. The FDA's guidance document requires the evaluation of four attributes to establish suitability, protection, compatibility, safety and performance drug delivery.

Protection

A container intended to provide protection from light, or offered as a light – resistant container must meet the requirements of the USP (661) Light Transmission Test. The ability of a container to protect against moisture can be ascertained by performing the USP (661) Water Vapour Permeation

Test. Integrity of the container can be evaluated using dye penetration test, microbial ingress test, gas detection test, etc.

Compatibility

Compatibility components will not interact with the dosage form and may not show leaching. Regular screening of the product using Liquid Chromatography (LC)/ Mass Spectrometry (MS), GC/MS, etc. would detect and quantify the elements. Other changes such as pH shift, precipitation, discoloration, which may cause degradation of the drug, should also be evaluated.

Safety

Packaging components should be constructed of materials that will not leach harmful or undesirable amounts of substances. This is difficult to determine. An extraction study with sample including preparation of sample, followed by incubation in solvents at well – defined and well – controlled times and temperatures is carries out. The USP Physiochemical tests (mentioned earlier) for extraction are relevant, sensitive and inexpensive.

Elastomeric container closure used in pharmaceutical, for injections, fall into this test class. Two extraction media are used, purified water and isopropyl alcohol. The extraction is performed at a specific temperature and time period. The tests performed are; turbidity pH change, reducing agent, heavy metals, total extractable zinc, ammonium etc.

Performance

The container / closure system should have the ability to function in an intended manner. It should function in a manner to improve patient compliance. Also it should have the ability to deliver the right amount or the right rate e.g. the prefilled syringes, transdermal patches, dropper or spray bottles, metered dose inhalers (MDI).

CHAPTER 8

Defects in Packaging

Summary

8.1 Introduction

Care should be taken to avoid unwanted defects in packaging. The common defects which may appear into the packaging includes: Cracked/ chipped bottle, Fracture in Can or body, Missing liner in closure, fracture in pouch, Heat seal is too narrow, Cartons out of square, Stones or air bubbles in glass bottle, uneven outer surface of bottle, small dents and scratches on case, printing out of register, washboard appearance on corrugated fiber board. Defects may be categorized as under:

Class A- Critical defects

Class B- Major Defects

Class C- Minor defects

8.2 Defects in Pharmaceutical Containers

The common defects, which may occur in most of containers like vials, ampoules, pouches, blisters, I.V. bags etc. include:

- Lack of heat seal - incomplete heat seal
- Delaminations, channel voids and contaminations
- Inclusions of product or foreign materials in the seal area
- Misplaced lids/tops/closures or crimp seals
- Invisible defects/leaks
- Holes or crack defects in empty vials/ampoules

8.3 Causes for Defects

1. Joint or material failures are potential problem areas.
2. If the packaging material is not sufficiently strong, then failure can occur as a collapse, rupture, split, tear or crack.
3. Closures can be another trouble spot

4. One of the biggest sources of supermarket customer complaints is badly applied caps on some pharmaceutical packages.
5. Packaging machines often run out of gas, or the machine controls are not set correctly, thereby producing thousands of out-of-specification packages.

8.4 Systems for Minimising Defects

There is a increasing demand for materials-testing equipment as a result of the drive to reduce packaging defects. There's a lot of pressure on the packaging industry to reduce packaging defects. Obviously that can be done through packaging design.

8.1 Introduction

Guaranteeing packaging integrity and eliminating defects such as leaky seals, faulty closures and incorrect pack-product combinations has never been more important. "In the enormously competitive retail market, as much consideration must be given to producing consistent, structurally sound and defect-free packaging as is given to pack presentation.

The drivers for this are numerous. Firstly, brand owners are looking to protect their most valuable asset – their brand. Litigation resulting from packaging defects makes press headlines and requires costly product recalls. Secondly, more focus is being placed on Hazard Analysis Critical Control Point (HACCP) and risk management. Finally, the contract-packing market is highly competitive and each producer is looking for an edge to differentiate itself.

This 'necessity' of consistently delivering flawless packaging has spawned a battery of integrity testing and inspection equipment, from leak detectors and torque testers to vision systems and materials testing.

When a perfect package is made it could be set to posses certain favorable qualities which include: containers should open and close easily, it must be free from scoffs and scratches, labelling graphics should be clear, it should be leak proof, directions on labels should be as per rules.

Care should be taken to avoid unwanted defects in packaging. The common defects which may appear into the packaging includes: Cracked/ chipped bottle, Fracture in Can or body, Missing liner in closure, fracture in pouch, Heat seal is too narrow, Cartons out of square, Stones or air bubbles in glass bottle, uneven outer surface of bottle, small dents and scratches on case, printing out of register, washboard appearance on corrugated fiber board.

Packaging defects may be categorized as:

1. Defects in containers
2. Defects in closures
3. Defects on labels
4. Defects on secondary packaging materials

Other way defects may be categorized as under:

Class A- Critical defects

Class B- Major Defects

Class C- Minor defects

(a) **Critical Defects**

Defects, which prevent container or package from performing its mini functions of protecting the contents, are referred as critical defects. It also prevents from compliance of regulatory region that causes hazards in handling or disposal. In case of critical defects container has to be rejected. Some examples of critical defects are : Cracked/ chipped bottle, Fracture in Can or body, Missing liner in closure, fracture in pouch.

(b) **Major Defects**

These defects have bottle like dysfunction. It reduces the identity by virtue of graphic defects. Some examples of major defects are : Heat seal is too narrow, Cartons out of square.

(c) **Minor defects**

Defects that damage the appearance but not the function of package are called minor defects. Some examples of minor defects are: Stones or air bubbles in glass bottle, uneven outer surface of bottle, small dents and scratches on case, printing out of register, washboard appearance on corrugated fiber board

Acceptance criteria for defects

Class A defects make a package unacceptable on subjective matter. It has been decided whether container is accepted or rejected. Class B defects has demerits of three points and class C has demerits of two points.

We accept a container of total ten points and disqualify the container on the basis of attributes and variables. Attributes are of yes/no type. For example, a container leaks or does not leak and closure is properly formed. Variable have a range hence specification gives acceptable range like bottle weight. If it comes under, lower limit then glass is too fragile, above the higher limit it will be too heavy.

8.2 Defects in Pharmaceutical Containers

The common defects, which may occur in most of containers like vials, ampoules, pouches, blisters, I.V. bags etc. include:

- Lack of heat seal - incomplete heat seal
- Delaminations, channel voids and contaminations
- Inclusions of product or foreign materials in the seal area
- Misplaced lids/tops/closures or crimp seals
- Invisible defects/leaks
- Holes or crack defects in empty vials/ampoules

8.3 Causes for Defects

1. Joint or material failures are potential problem areas. From ultrasonic welds to adhesive bonds, joined parts will always pose a risk of failure. Films and foils sealed over polystyrene containers, foil blister packs, crimped toothpaste ends, drink cartons and sauce sachets are all typical examples.
2. Also, if the packaging material is not sufficiently strong, then failure can occur as a collapse, rupture, split, tear or crack.
3. Closures can be another trouble spot: Screw closures must be applied with precisely the correct level of torque. Too much may damage a waterproof cap lining and compromise the seal, whereas too little may leave the cap insufficiently tight to create a hermetic seal.
4. One of the biggest sources of supermarket customer complaints is badly applied caps on milk bottles and some pharmaceutical packages. For this application, some manufaturers supplies charge-coupled device (CCD) vision systems to check bottles after capping but before heat sealing.
5. Packaging machines often run out of gas, or the machine controls are not set correctly, thereby producing thousands of out-of-specification packages.

Some common packaging and their critical, major and minor defects are given in the Table 8.1.

Table 8.1 Some common packaging and their critical, major and minor defects.

Sr. No.	Package Type	Critical Defects	Major Defects	Minor Defects
1.	Glass container	-Cracks and penetrating checks -Light side walls -Broken/ chip finish -Contaminations with insects	-Nonfunctioning stretching bead liners in excess of 1/8 inch on each side	-Stones and heavy appearance followed by rough molding lines and uneven outer surface.
2.	Labels	-Wrong copy/ size. -In stick together -Tear and Holes -Smeared or illegal	-Rough weight edges -Stretched or below standard scoff resistance	-Loosely bonded. Improper identification on label

Table 8.1 contd...

Sr. No.	Package Type	Critical Defects	Major Defects	Minor Defects
3.	Labels for Shipper	-Dimensions outside the limit break	-Bands having score lines less than ten. -Moisture content below 5% or above 30%.	-Deviate less than one inch -Stretched on ends and interior.
4.	Metallic Packages	-Leakage on seals -Crack of any size on body or ends -Missing or incomplete lining	-Dent size is more than a inch long that affects closing efficacy.	-Sealing compound below minimum wt. -Solder ends open
5.	Blister packages	-Crack or holes on blister -Wall thickness less than 0.005 inch -Draw to shallow for contaminant of the product.	-After washing it is not removed -Embedded foreign matter. -Discoloration and other visual defects. -Failure to meet scratch resistance test.	-Rough weight flanges. -Mould lines and drawing lines.
6.	Plastic Bags	-Unsecured mounting -Holes, gaps, tears, inside seals	-Contamination of dust, grease and other foreign matter that can't be removed with normal washing.	
7.	Plastic Bottles	-Crack or break of bottle -Less than 0.005 inch in thickness -Rough wt out at finish leading to leakage through closure	-Liner encoding specifications below wt. -Bottle crack are of pinch or adhering from base.	-Blotting or incomplete printing -Wall thickness less than 0.01 inch so as to loose the opacity. -Inadequate pigment mixing.
8.	Closures	-Partially filled cavity. -Tight closures outside the dimension that permits stress cracking.	-Faults related to consumer -Parting liner other than closure surface. -Rough or orange pill surface.	-Trimming is improper. -Improper identification of cases.

Table 8.1 contd...

Sr. No.	Package Type	Critical Defects	Major Defects	Minor Defects
9.	Cartons	-Dimensions outside the tolerance limit - Tear , holes - Mis-registration of scores	- Ink spreading -Scoffs surface -Moisture content below 5% or above 30%. -Flaps unlocked.	- Light blotchy graphics - Uneven strains on outer surface.
10.	Aluminum Foil	-Damage core -Low gram age thickness -Missing sealing layer	-Pinholes -Width variation. -Poor rub resistance -Blotted printing	-Scratches on the lamination. -Bead over surface of foil. -Sealing is not proper.

8.4 Systems for Minimising Defects

There is a increasing demand for materials-testing equipment as a result of the drive to reduce packaging defects. There's a lot of pressure on the packaging industry to reduce packaging defects. Obviously that can be done through packaging design.

Once the preserve of pharmaceuticals, vision inspection is becoming an increasingly popular technique for checking fill level, cap position, tamper band presence, label presence and placement and data in a range of industries. In our labelling lines we are being asked to check for label presence, correct label information, both in text and barcodes, and even the accuracy of registration of the labels. Manufaturer can incor-porate camera vision systems, which will recognise shapes, sizes and colours of products, with operator warning systems, both visual and audible. Many of these innovations have been developed within the pharmaceutical sector, where the integrity and security of products is paramount, but their use has been readily absorbed into other areas where product protection is also of the essence. Some of the major machine precautions taken into consideration to reduce defects are described below:

- Joint-strength test: a generic tensile test to destruction on two mating components
- Top-load test : a compressive test performed on containers to assess resistance to topload
- Pierce-force testing of foils and films.
- Trouser-tear tests: recording the force required to commence and propagate a tear

- Simple tensile or compressive tests on packaging material.
- For peelable lids on containers such as yoghurt pots and daily-dose drink bottles : it offers equipment capable of testing peel strength
- Testing machines like texture analyser: records the force required to open peel-back lids.
- The caps are placed on the bottles, tightened and then heat sealed: if the cap is not applied absolutely square, the seal will not be effective.
- Vision inspection outside the pharmaceutical industry: Machine Vision Technology (MVT) designed a vision system that reads the barcode on each can label and matches it with the product code on the bottom of the can, at speeds of up to 1,200 cans per minute.

CHAPTER 9

Stability of Packages

Summary

9.1 Introduction

Medicinal and Pharmaceutical products are subject to a variety of degradation pathways that compromise drug safety and shelf-life. By far the greatest degradation is caused by hydrolysis and oxidation. However, other mechanisms include racemization, photodegradation, elimination, and complexation. A variety of packaging components are available to help maintain pharmaceutical integrity; however, determining the optimal solution requires a calculated analysis that joins the expertise of packaging engineers and pharmaceutical formulation chemists.

9.2 Measures to Improve Stability

When maximizing shelf life through such formulation approaches, the formulator wants assurance that the packaging will not shorten the shelf life of the product. Thus a careful analysis of the relation between package properties and shelf life is important. Prediction of shelf life of a packaged product is a topic of pharmaceutical packaging that has been actively and innovatively pursued for at least two reasons. First, the time-honored approach to determining shelf life employs accelerated stability studies at elevated temperatures, along with parallel confirmatory real-time studies.

- Modeling Moisture and Oxygen Ingression
- Examining the Whole Process
- Consider Changing Protection Strategies

9.3 Pharmaceutical Package Stability Testing

As regulators increasingly look for proof of pharmaceutical product stability throughout the supply chain, it is critical for manufacturers to ensure that pharmaceuticals maintain stable temperatures and endure handling from warehouse - to transit - to shelf.

Following are the types of pharmaceutical package testing as per PDA Technical Report

- Thermal Performance Testing
- Stability Testing
- Distribution Simulation

9.1 Introduction

Medicinal and Pharmaceutical products are subject to a variety of degradation pathways that compromise drug safety and shelf-life. By far the greatest degradation is caused by hydrolysis and oxidation. However, other mechanisms include racemization, photodegradation, elimination, and complexation. A variety of packaging components are available to help maintain pharmaceutical integrity; however, determining the optimal solution requires a calculated analysis that joins the expertise of packaging engineers and pharmaceutical formulation chemists.

Assigning the stability, it is suitable to first assess the principle by which adequate protection is judged, namely expiry. The period of time over which the specific dosage form will adequately preserve its potency, purity, bioavailability, and relevant physical properties under usual storage conditions is known as Expiry. Frequently referred to as shelf life, expiry varies with the specific product and often equates to the time for a product to degrade from 100% to 90% of its labeled potency i.e. there becomes 10% degradation in potency. A broader definition is the time to reach 90% from a larger initial potency.

Barrier properties of typical plastic bottle materials.

Property	**Permeability constant [(cm^3 mm)/(sec cm^{2}cm Hg)]**			
Permeating gas molecule	H_2O	O_2	CO_2	N_2
Temperature (^{0}C)	25	30	30	30
Type of plastic				
Low-density polyethylene	800	55	352	19
High-density polyethylene	130	10.6	35	2.7
Polypropylene	680	23	92	4
Polyvinyl chloride (unplasticized)	1,560	1.2	10	0.4
Polyvinylidene dichloride (Saran)	14	0.053	0.29	0.0094
Polystyrene	12,000	11	88	2.9
Polyester terephthalate	1,300	0.22	1.53	0.05

Sorbents such as desiccants and oxygen absorbers represent a class of active packaging components that can be used to guard against the effects of degradation. These ensure the integrity of packaged pharmaceuticals. Sorbents are considered "active" because they respond to changes in the headspace of packaging relative to outside conditions. Packaged sorbents are manufactured from silica gel, molecular sieve, and a variety of oxygen-absorbing compounds.

Active sorbents have high adsorbing capacity and take up little space. When incorporated successfully into a package, they can significantly reduce the rate of degradation, resulting in an improvement in pharmaceutical quality, safety, shelf-life, stability, and usability.

Sorbents come in different delivery formats, including canisters and packets, as well as a new compressed density format. Determining the correct active formula requires manufacturers to consider a variety of factors.

9.2 Measures to Improve Stability

When maximizing shelf life through such formulation approaches, the formulator wants assurance that the packaging will not shorten the shelf life of the product. Thus a careful analysis of the relation between package properties and shelf life is important. Prediction of shelf life of a packaged product is a topic of pharmaceutical packaging that has been actively and innovatively pursued for at least two reasons. First, the time-honored approach to determining shelf life employs accelerated stability studies at elevated temperatures, along with parallel confirmatory real-time studies. Such an approach is costly in materials, possesses higher risk because of inadequate background information, and perhaps more important, is time-consuming. Second, it is generally felt that drug products are frequently "overpackaged", that is, the packaged offers protection of the product significantly beyond that required for the desired expiration date. This is a natural outcome when inadequate background information exists and an attempt to minimize the downside risk is made. This process will again increase the total cost of the product.

The objective of package stability studies, then, is to match package performance with needed protection, using cost-effective package materials, and to do so in the shortest time possible. The various shelf life prediction models that have been recently developed are designed to address these concerns. To develop a reasonably accurate shelf life model, it is necessary to have some knowledge of the product, the package, and the environment. Synthesizing information from these three sources and obtaining shelf life predictions began primarily in the field of food science.

More recently, others have amplified and applied these earlier theories to obtain very realistic estimates of shelf life.

A major driving force for attempting to predict shelf life is the potential inapplicability of accelerated testing for a number of drugs products. The rules and assumptions for using Arrhenius equation on the temperature dependence of a degradation reaction often break down for a solid dosage form. Consider the situation of a coated tablet in a plastic bottle, in which the various polymeric materials in both product and package might each pass through temperature-dependent transitions (e.g., glass transitions or thermal melts). Such thermal effects causes changes in the actual physical state of the polymers and, as a result, seriously impair the reliability and predictability of the accelerated test method. These kinds of transitions render it improper to use the *Arrhenius equation* for rate constant versus temperature data because the mechanism of degradation under the accelerated conditions is different from that under ambient storage conditions. In the extreme, a similar argument can be made for even the pure solid phase of drug, when as has often been discovered, the pure drug alone may exit in any of several temperature-dependent morphological forms. Then, not only does the crystalline structural form change, but, also, conceivably, the rate and mechanism of degradation.

Modeling Moisture and Oxygen Ingression

To optimize package protection for solid-dose formulations, pharmaceutical manufacturers need to employ moisture and oxygen ingress modeling techniques to analyze the rates of degradation for a given pharmaceutical. The rationale behind modeling is based on a number of factors. Oxidative changes can cause a loss of pharmaceutical potency after several weeks, whereas two to three years of shelf-life may be required depending on distribution channels. Modcling also determines steady-state levels of oxygen within bottles, moisture permeation across bottle walls and material permeability.

Ingress modeling provides a calculated moisture or oxygen ingress value using established Moisture Vapor Transmission (MVTR) and Oxygen Transfer (OTR) rates. MVTR and OTR rates are applied to the proposed package and, with the inclusion of other sources of moisture or oxygen, are used to determine the appropriate amount of sorbent material to be used.

Desiccant and oxygen absorber requirement calculations rely on interdependent dynamic modeling, which judges the interaction of package permeability with the ad-sorption and de-sorption properties of the pharmaceutical formulation.

Ingress modeling results and related sorbent product recommendations allow packaging engineers and formulation chemists to quickly understand what their theoretical sorbent requirement will be for a proposed package. It also allows them to demonstrate these findings using a small-scale, "proof-of-concept" accelerated stability test.

Ingress modeling requires specialized equipment as well as skilled analysis by people experienced in packaging challenges and analytical chemistry. It is crucial to find a packaging solutions provider proficient in both of these areas of expertise.

Examining the Whole Process

When shaping a degradation-prevention program, it is also critical that manufacturers consider the whole manufacturing process. What might work in one part of the operation may not work at another end - solutions determined upstream may have unintended consequences downstream.

For example, ingress modeling may propose a two-gram unit sorbent that does not fit into the neck of a proposed pharmaceutical bottle. Another critical consideration is the type of automation found in a packaging line. A sorbent format may not work at high speeds with certain automated equipment. Other times, a "double drop" of sorbents may be required, depending on the bottle dimensions, or the pharmaceutical.

An active packaging solution must consider all parts of the manufacturing process, from the pharmaceutical formulation to the packaging environment and to the distribution chain. When developing the optimum package protection, the best approach to take is a multidisciplinary one that considers the knowledge and experience of packaging engineers, formulation chemists, analytical chemists, and the sorbent supplier.

Consider Changing Protection Strategies

Pharmaceutical manufacturers often assume that it is time consuming and expensive to change strategies once an active packaging protocol has been established and approved. This is not typically true. In fact, changing course may significantly streamline operations and improve product quality, and often can be more costeffective. The most time-consuming aspect of switching strategies is finding out what needs to be done. Current FDA guidance has made changing desiccant, or oxygen absorber, formats a relatively simple procedure. Often it is a simple matter of an annual reporting change. It is important; however, to work with a packaging partner who understands the regulatory roadmap. All products should be 21 CFR-compliant and supported by a Type III Drug Master File (DMF) on file with FDA.

Protecting pharmaceuticals from degradation is important to ensure commercial success. When used appropriately, active packaging is an important part of the complete solution. As with any evolving technology, a change may appear daunting but can, in fact, improve operations. The bottom line is to consider the whole gamut of factors involved in pharmaceutical degradation, and every aspect of upstream and downstream processing. Calculating these factors with a modeling system, supported by a knowledgeable partner, can lead to a more shelf-stable, and safer, product.

9.3 Pharmaceutical Package Stability Testing

As regulators increasingly look for proof of pharmaceutical product stability throughout the supply chain, it is critical for manufacturers to ensure that pharmaceuticals maintain stable temperatures and endure handling from warehouse - to transit - to shelf.

Following are the types of pharmaceutical package testing as per PDA Technical Report

Thermal Performance Testing

From highly controlled cold chain package testing facilities, located in Minnesota and California, tests thermal packaging for pharmaceutical companies.

Stability Testing

Over a 72–120 hour timeframe, it conducts a stability test and develops a specification designed to inform pharmaceutical package testing clients of conditions that will push a thermal container out of the 2-8 centigrade (an indication that contents may lose active ingredients or even spoil) desired shipping and storage temperature.

Distribution Simulation

Experts in package testing engineers test EPS materials and hard foam packaging. They also tests packaging inside of pharmaceutical shipping containers including gel packs, glass vials, bottles and other types of interior pharmaceutical packaging. Dynamic distribution and drop testing validation of pharmaceutical packaging is also tested.

Conversely, an alternative shelf life prediction approach characterized the properties of the components of the system under ambient conditions. By measuring the time rate of change of these properties with sufficient precision, a mathematical model was applied, and a shelf life was predicted. In the traditional stability study, the rate of degradation of the active ingredient is measured by a stability-indicating method, which usually involves the measurement of a small change in a relatively large quantity. For example, when following the degradation of 100 mg of active drug in a tablet, 1% degradation represents a 1.0 mg loss in the active agent. The loss will correspond to a 1% reduction in the measured property being used to quantitate the active drug. Furthermore, variability of content uniformity, the imprecision and inaccuracies of the assay method, and other sources of error may require that a number of tablets be assayed to identify whether or not a 1% decrease in potency is statistical significant. In contrast, it showed that by following the appearance of a degradation product, sensitivity is vastly improved. In the foregoing example, the appearance of the degradation product corresponding to the 1% loss of active drug represents a 100% increase (from 0 to 1 mg) in the parameter quantitating a degradation product. This increase in signal/ noise ratio may allow the investigator to eliminate working under accelerated conditions and, instead, determine degradation rates in real-time under normal conditions.

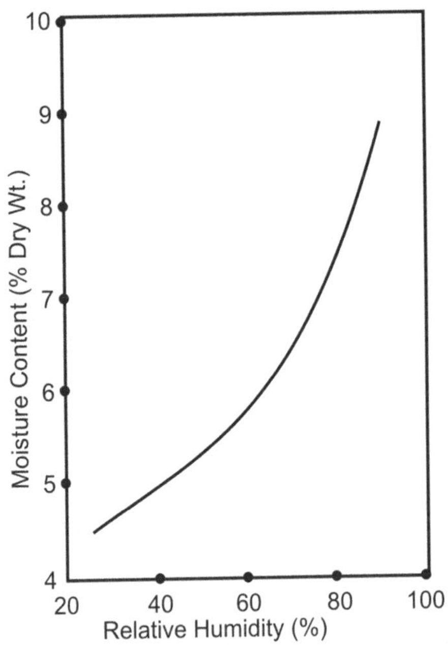

Fig. 9.1 Relation of equilibrium moisture content and relative humidity.

Some workers took a different approach in shelf life prediction and employed the food-packaging concepts to characterize the behavior of drug products at constant temperature as a function of equilibrium moisture content (EMC). Results were obtained based on each of three fundamental considerations that relate to the stability of the drug product; that is, the product, the package, and the environment. As shown in Fig. 9.1 and 9.2.

The EMC of the product was determined as a function of relative humidity. Then a physical or chemical property related to dosage form stability was measured as a function of relative humidity or moisture content. Give these data, critical moisture content was defined as that beyond which the drug product no longer met specification and, therefore; had reached expiration. Separately, a third set of results, the package moisture transmission rate, was characterized under known environmental conditions (e.g. 25 °C and 50% RH). Some experimental observation are presented in Fig, 9.3 and Fig. 9.4.

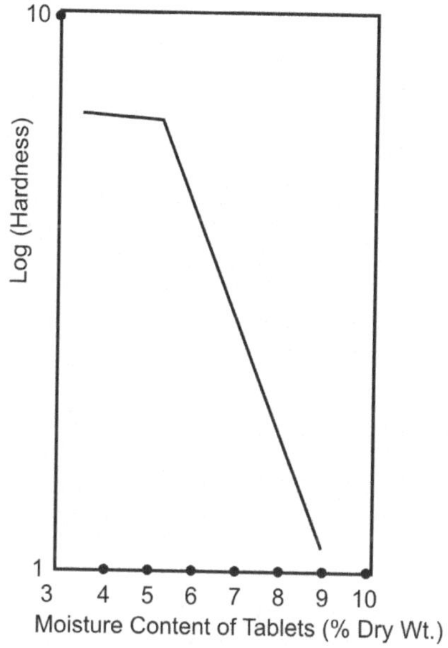

Fig. 9.2 Relation of drug properties (Hardness, assay etc.) and moisture content.

The most complete model must correlate package performance to the degradation rate of the drug product or the rate of change of other critical dosage form properties that relate to stability.

This would then lead to a dynamic model wherein the kinetics of degradation, in terms of the reaction rate constants and package permeation rates, are correlated.

This analysis does not render the existing models unusable, because there is situation during which the active drug is extremely inert, but the physicomechanical properties of the dosage form change with, and are dependent on, moisture content. Under such circumstances this approach appears to work well to predict shelf life.

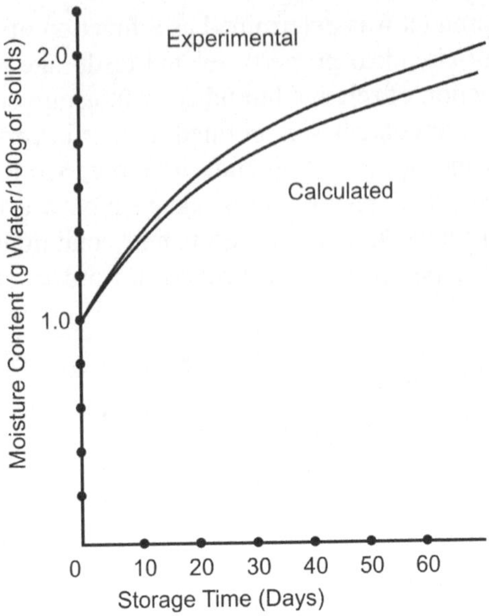

Fig. 9.3 Prediction verses observation of Moisture content upon storage at 22°C and 73%RH of a multivitamin tablet.

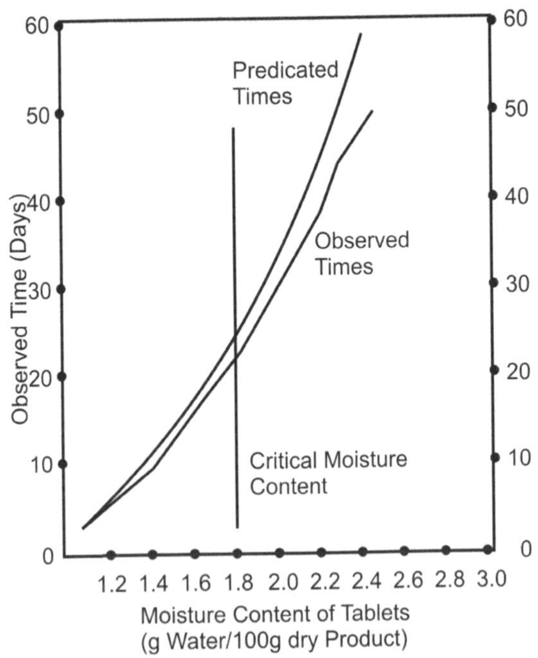

Fig. 9.4 Prediction verses observed time to reach Equilibrium Moisture content upon storage at 22°C and 73%RH of a multivitamin tablet.

CHAPTER 10

Ancillary Materials Used in Packaging

Summary

10.1 Introduction

Labels are slips made generally of paper and are attached to a container bearing name, description and other details regarding the product. Another feature being introduced into the pharmaceutical label is bar coding. EAN (European Article Numbering) bar coding has become an integral part of label for products competing in the global market.

Ancillary materials used in pharmaceutical packaging are as follows

1. Paper and Boards
2. Wood and Jute
3. Inserts and Leaflets
4. Composite containers may be used as tubes for protection of small powder packs

10.2 Manufacturing of Papers and Boards

Manufacturing of papers or boards includes two basic processes

1. Pulping
2. Machine conversion into paper or board

10.3 Application of Various types of Papers in Pharmaceuticals

Specific reference is made to four main types of Paper labels

1. Pain with adhesive added
2. Pre-gummed
3. Heat Sensitive
4. Pressure Sensitive

Use of papers in Leaflets

Paper liners, linings and lamination containing paper

10.4 Boards

There are various types of corrugated fiberboards

(a) Single wall i.e. one sheet of filter paper between two liners.

(b) Double walled i.e. two sheet of corrugated paper between three liners.

(c) Triple walled- three sheet of filter paper between four liners.

10.5 Testing of Paper and Board

Specific tests for Cartons

1. Compression test.
2. Carton opening force.

Hold the flat carton as delivered by its crease between thumb first finger and press. The carton should spring open into the square position without a need for unreasonable force.

1. Coefficient of friction

   Both the static and kinetic coefficient of friction is determined by sliding the specimen over itself under specific test conditions.

10.1 Introduction

Ancillary materials are widely used in pharmaceutical industry to confirm the packing of completed products. It is important to use these materials to get ultimate packs of primary packages.

Paper is widely used due to the advantages like low cost, wide availability, non toxic origin, and availability in variety of range, ease of printing, coating, lamination, opening, use, and modification by addition of additives, good rigidity, strength, biodegradability and disposing.

It suffers from some drawbacks of being moisture responsive, need alteration in permeability to moister and gases, is porous and is unable to form a barrier. However, advantages of paper outweigh its limitations and it is used widely for formation of labels, leaflets, etc. Strength of paper depends on its GSM (gram per square meter). This serves as an important parameter for its quality control.

Paper and board may be used for applications such as wrapping materials. Labels, patient package inserts (PPI), bags and sacks, corrugated boxes, cartons, shipment containers, gummed tape, fitment for cases, paper liners, lining and lamination. Without these, packaging of products would be more expensive and difficult.

Labels are slips made generally of paper and are attached to a container bearing name, description and other details regarding the product. Another feature being introduced into the pharmaceutical label is bar coding. EAN (European Article Numbering) bar coding has become an integral part of label for products competing in the global market. This coding is a facilitation tool to track the products effectively at any point of time to ensure effective supply chain management and speedy product recall in need There is an increasing trend of including information leaflets into products. Patient package inserts (PPI) are such leaflets, which are inserted into the product package. These contain non-technical explanation informing patients about the medicine in the package.

Ancillary materials used in pharmaceutical packaging are as follows

1. Paper and Boards
2. Wood and Jute
3. Inserts and Leaflets
4. Composite containers may be used as tubes for protection of small powder packs

Paper board and other cellulosic material have a major future in pharmaceutical industry. These are renewable and economical alternative to petroleum-based materials. It may be used in the following Pharma packaging applications

1. Labels and Leaflets
2. Wrapping materials
3. Bags and Sacks
4. Collapsible and Rigid Cartons and Boxes
5. Shipping and Transit outers
6. Gum Tapes
7. Filaments and Cases
8. Composite tubes and drums
9. Molded pulp board containers
10. Paper liners, linings and Lamination

10.2 Manufacturing of Papers and Boards

The history of paper making in particular is long. It has always used forms of cellulose fiber, but began to use wood pulp in mid nineteenth century. Boards were introduced in the same period as hand made construction to help, protect and transport products. Corrugated boards is a recent introduction, starting in the early 1870's and became widely commercially available in the late 1890's.

Manufacturing of papers or boards includes two basic processes

1. Pulping
2. Machine conversion into paper or board

10.2.1 Pulping

The treatment of fibers (pulp) is the major influence on the properties and cost of the paper of board produced. The objective in pulping is to 'tease' on undamaged fibers from the mass of wood, so that these fibers can be recovered into smooth paper or board required.

There are three major processes which reduce raw materials

1. Mechanical pulping
2. Chemical pulping
3. Semi chemical or Combination pulping.

The wood pulp is supplied from two most popular sources.

1. Managed softwood forests
2. Recycled fibers.

Most label papers are chemical pulp with additives. Most box boards are layers of mechanical or semi-chemical pulps on the face and some times the back facing.

10.2.2 Bleaching

These are grades probably used for white paper and white plies of the board.

Chemical pulp made from sulphite or soda process

↓

Mixed with hydrogen chloride or hypochlorite dissolved in water

↓

Removal of colored residues in the cellulose fibers

↓

Chemical pulp is bleached.

10.2.3 Beating and Refining

Beating is the batch process where the pulp suspension is recycled through specially designed vessels which shortens and softens the fibers.

Refining is a continuous process where by the pulp suspension is pumped between a static outer and tapering rotor.

10.2.4 Mixing Stage

The pulp is vigorously stirred in a hydro pulper with copious amount of water and the additives are added to make the final mixture.

10.2.5 Additives

These are added to the beaten or refined pulp to produce the final form of the paper.

The typical additives are as follows:

1. Loading and Fillers : To improve opacity/ brightness.
2. Coloring materials.
3. Sizing materials : To reduce the penetration of water and inks.
4. Binding agents : To increase strength.
5. Gums.

6. Antifoam agents.
7. Wet strength resins.
8. Optical brightening agents (OBAS).

10.2.6 Machine conversion into paper (Fourdrinier System)

This is the most popular system still used today. The machine starts at the 'wet end' and finishes at the 'dry end'.

10.2.7 Instrumentation

1. A stuff chest
2. A Brest box or head box
3. A slice
4. Fourdrinier wire with vacuum boxes
5. Presses
6. Driers
7. Machine glaze drier
8. Calendar stacks
9. Reeling

Suspension of pulp (90% water)
including all the additives

↓

Pumped into the stuff chest

↓

Led down into the 'head box'

↓

Fed into 'head box' through variable 'slice'

↓

Final sites over a fast moving fine
wire mesh i.e. Fourdrinier wire

↓

The water from suspension drains
away leaving the mass of solid on the wire

↓

Wires induces the fibers to align themselves
Preferentially to the machine direction

↓

Producing the 'grain' of machine made paper.

10.2.8 Finishing process

This consists of the additional treatment which is carried on the material after the paper making process. E.g. Surface sizing may be performed by applying a solution of gelatin together with other chemicals to the paper.

The advantage is that it improves water resistance and printing properties and is called 'tub-sizing'.

10.2.9 Other surface treatments

These includes:-

1. Cloaking
2. Impregnations
3. Laminations

These are mainly aimed for reducing moisture, gas penetration or for creating a heat sealing capabilitics.

1. Dry waxing process.
2. Wet waxing process.

The most common type of paper used in pharmaceutical packaging are as follows

1. 'Kraft paper'- Used as an outer facing for corrugated board, solid Board, spirally wound bags and fiberboard drums.
2. 'Uncoated paper'- Usually from high grade pulp source used in thin calipers for small labels and leaflets.
3. 'White wood free paper'- For lamination which is usually one side coated paper that has been super calendared to make the (coated) surface more permeable.
4. 'Vegetable parchment paper' is made by a process of treating the absorbent paper with sulphuric acid which enhances the wet strength of the paper.

10.2.10 Machine conversion into Boards

Boards are rigid and folding boxboards, solid and corrugated fireboard, fire drums, and components of composite containers.

10.2.11 Vat or Cylinder machine

The machine starts at the 'wet end' (1-4) and finishes at the 'dry end' (5-9) just like Fourdrinier machine.

Steps involved:-

1. Pulp preparation

2. Make up tanks with different pulps
3. Vats with a constant level of pulp
4. Vacuum applied cylinders in Vats
5. Presses
6. Dryers
7. Machine glaze driers
8. Cylinder stacks
9. Reeling

Twin wire machines are Fourdrinier type processes which are drained between two forming fabrics or wires in a horizontal plane.

There are also multi-wired Fourdrinier types of machine in operation today.

10.3 Application of Various types of Papers in Pharmaceuticals

10.3.1 Use of papers in labels and Labelling

Specific reference is made to four main types of Paper labels

1. Plain with adhesive added
2. Pre-gummed
3. Heat Sensitive
4. Pressure Sensitive

Plain with adhesive added

These are applied after the addition of adhesive. Type of adhesive used depends on the surface of the item to be labeled. Labelling to plastic surfaces requires the use of specialized adhesive which may be based on latex or synthetic resins e.g. polyvinyl acetate.

In certain instance pretreatment of the plastic with flaming or carbon discharge can aid adhesion. Dextrin is the most widely used adhesive involving different levels of solid contents.

Pre-gummed Labels

Paper is pre-coated with dextrin or gum Arabic, as only water is required to activate the adhesive. The labels are cleaned to apply and a complete labelling operation can be carried out by one operator. They are ideally suited for small runs or intermittent production, such as partial labels. They provide good adhesion to paper board, glass but as the range of adhesive type is very limited they lack flexibility in terms of adhering to the wide range of surfaces

Heat Sensitive or Thermoplastic Adhesive Based Labels.

They are activated by heat and do not rely on conventional gums or water as a wetting agent but they adheres to the wide range of materials including metals, plastics and varnished coated and printed surfaces

Two types of heat sensitive labels are in use viz. Instant tack and Delayed Tack.

Pressure Sensitive Labels

Pressure Sensitive Labels are removed from backing paper and then applied by pressure. Ideally they should be removed by turning the backing paper over a right angle so that the labels come off straight. They consist of suitable label paper coated on the reverse side with a permanently tacky adhesive which in contact with the back paper that protects it prior to use.

10.3.2 Use of papers in Leaflets

Paper leaflets are broadly of four types as

Cut sheet

It is usually printed on both the sides, delivered as blocks of cut sheet and folded on cartoning machine

Reel Fed

These are like cut sheet labels but with no backing paper.

Pre-foldcd

These are delivered as bundles and these need to be broad banded not with elastic band around them or contained in plastic cartridges and fed via a hopper system direct to the cartoning machine.

Combined labels/ Leaflets

They are delivered as a thick pressure sensitive label either reel fed or block cut containing a fold out portion which is the leaflet itself.

10.3.3 Paper liners, linings and lamination containing paper

Papers of various grades are used as liners inside bags made of another material. Glassine has been used as linings for the inside of plastic, metal or fiberboard bags. Paper is used extensively in laminations where it is bonded, usually by heat and or adhesive to other films, notably to plastics and Aluminium, eg. LDPE/adhesive/Aluminium/adhesive calendared paper.

These types of laminations can be used for fill and seal of materials.

10.4 Boards

The common types of boards which are used are

1. **Millboard**

   These can be made on a single cylinder or Vat machine, where a moist web of thick paper is built up in plies and then milled between heavy rollers.

2. **Clipboards**

   These use post industrial waste and some post consumer waste in their middle plies.

   They can be faced with better quality facing materials and may or may not have a quality backing used for low grade small boxes and cartons.

3. **Recycled boards**

   They consist of top coating, top liners, under liners, middle plies and back liners.

4. **Virgin boards**

   They consist of top coating, base coating, chemical pulp and mechanical pulp.

5. **Strawboards (Lined or unlined)**

   They are used for rigid boxes e.g. solid energy boxes where the slight odour, high chlorine content are of little importance. Fiber is provided by strops and is made on a usual board machine. Lining is achieved by pasting paper with the desired type to both sides on a continuous process machine. It is quiet rigid but has limited use in pharmaceuticals because of the slight odour.

6. **Solid fiberboards**

   This is two or six layers of recycled paperboard which is treated with an adhesive between the layers and press laminated collectively. The outside layer can be 'Kraft paper' to get better the strength. It has good crush strength.

7. **Corrugated fiberboard**

   This is the most popular form of outer protection used in the industry today. It comprises of one or more sheets of fluted (Corrugated) paper protected by an adhesive to two or more liners. The paper used for the Corrugation is made up of recycled paper. Corrugated layer is impregnated with a stiffening agent usually starch or synthetic vinyls. It is frequently printed by flexography or inkjet which can weaken the board.

There are various types of corrugated fiberboards

(a) Single wall i.e. one sheet of filter paper between two liners.

(b) Double walled i.e. two sheet of corrugated paper between three liners.

(c) Triple walled- three sheet of filter paper between four liners.

10.5 Testing of Paper and Board

The reason for testing is to give information thereby ascertaining the acceptability of the test piece in comparison with a specification. Testing conditions have also to be defined, as paper and board are very susceptible to humidity and temperature changes. The test pieces for paper and board are in fact 'conditioned' so that the test can be carried out under 'standard' conditions there by giving fair comparison.

Those conditions are:

Temperature 23 °C ± 1 °C

Relative humidity 50 % ± 2 %

10.5.1 General tests for Paper and Boards

1. Sample conditioning for Paper and Board
2. Pre-test procedures for Paper and Board
3. Moisture content of Paper and Board
4. Density of Paper and Board
5. Methods for determining air permeability
6. Methods of test for the assessment of odour for packing materials used for food stuff.
7. Tensile strength both wet and dry.
8. Tear strength either across or along.
9. Burst strength both wet and dry.
10. Puncture resistance.
11. Stiffness of thick Paper and Board
12. Ply bond of Boards.
13. Crushability of Boards.
14. Cobb test.

15. Rub resistance.
16. Pick test
17. Roughness/Smoothness
18. Brightness
19. Opacity.
20. Dennison wax test
21. Wet burst strength
22. Wet tensile strength.
23. Ash in Paper and Board.
24. Detection and estimation of nitrogenous agents in Paper.
25. Ink absorbency.

10.5.2 Specific tests for Cartons

1. Compression test.
2. Carton opening force.

   Hold the flat carton as delivered by its crease between thumb first finger and press. The carton should spring open into the square position without a need for unreasonable force.
3. Coefficient of friction

   Both the static and kinetic coefficient of friction is determined by sliding the specimen over itself under specific test conditions.

CHAPTER 11

Sterilisation of Packaging

Summary

11.1 Introduction

The techniques related to sterilization of Pharmaceutical packaging materials is not stagnant, new techniques, recent improvements and modifications have been developed. All products labeled sterile must pass the sterility test having been subjected to an effective process of sterilization

11.2 Sterilization Methods

Having conventional a validated procedure, knowledge of its performance in regular use should, wherever possible, be gained by monitoring and suitably recording the physical and where relevant, chemical conditions achieved within the load in the chamber during each sterilizing cycle. Appropriate sterilization cycles can be established only on the strength of this information.

1. **Steam Sterilization (Heating in an Autoclave)**

   This process involving heating in an autoclave with saturated steam under pressure should be used whenever possible for aqueous preparations and for surgical materials. An aqueous product which is to be sterilized by this method is distributed into suitable containers which are than sealed so as to exclude micro-organisms.

2. **Dry Heat Sterilization**

   This process is suitable for heat-stable, non-aqueous products and powders. A preparation to be sterilized by dry heat is distributed in the final containers which are then either finally sealed or temporarily closed to exclude micro-organisms and then heated so as to ensure that the entire contents of each container are maintained for an effective combination of time and temperature to provide an adequate assurance of sterility.

3. **Gas sterilization**

   This method involves exposure to a sterilizing gas. The gas generally employed in gaseous sterilization is ethylene oxide of acceptable sterilizing quality. The process is difficult to control, the gas itself being toxic and, when mixed with certain proportions with air, explosive.

4. **Sterilization by Ionizing Radiation**

   Sterilization of drug substances, dosage forms, packaging materials and medical devices may be achieved by exposure in the final containers or packages to ionizing radiation in the form of gamma radiation from a suitable radio-isotopic source such as cobalt-60 or cesium-137 or of electrons energized by a suitable electron accelerator.

11.3 Factors Affecting Selection Sterilization Method

The first factor to be considered with medical packaging materials is the sterilization method. Many times the packager is committed to a certain method because of product restrictions, process economics, or equipment availability. If end-users are locked into autoclave sterilization, then heat-resistant materials must be used. Some medical products are incompatible with specific sterilization methods. For example, one would not sterilize liquids (such as filled IV bags) with ethylene oxide (EtO), as this process requires permeable packaging materials. For products that are terminally sterilized, the packager's choices are often significantly narrowed based on the sterilization method that will be used.

11.1 Introduction

Complete absence of viable microorganisms in any object would be deemed as sterile object. The objective of sterilization process is to remove or destroy all microorganisms in or on a preparation/material and to assure in this way the preparation is free of contagious hazards in all respect. However, this absolute explanation cannot be applied to an entire batch of finished pharmaceutical articles because of the limitations of testing. Total sterility cannot be practically established without complete destruction of microorganisms in every finished article. The techniques related to sterilization of Pharmaceutical packaging materials is not stagnant, new techniques, recent improvements and modifications have been developed. All products labeled sterile must pass the sterility test having been subjected to an effective process of sterilization. A sterilization process requires the validation of sterilizing equipments on following steps:

1. The sterilizing equipment has the capacity to operate within the required parameters.
2. The critical control equipment and instrumentation are capable of operating within the prescribed parameter for the equipment.
3. Replicate cycles representing the required operational range of the equipment and employing actual or simulated product.
4. The probability of microbial survival in the replicate processes completed is not greater than the prescribed limits.
5. Documentation of above step with complete validation protocols.

The validation programme for aseptic processing in which the components of the final dosage form are sterilized separately and the finished product is assembled in an aseptic manner is similar to the validation of a sterilization process. The validated and certified process should be revalidated periodically; however, the revalidation need not necessarily be as extensive as the original one.

It is generally accepted that terminally sterilized injectables preparations purporting to be sterile, when processed in the autoclave, attain a 10^{-6} microbial survivor probability, i.e. assurance of less than one chance in one million that viable micro-organisms are present, in the sterilized article or package form. With stable articles, the approach is to considerably exceed the critical time necessary to achieve the 10^{-6} microbial survivor probability (overkill). However, with an article where excessive heat exposure may have a damaging effect, it may not be feasible to employ the overkill approach. In such cases, the development of the sterilization cycle depends heavily on knowledge of the microbial burden or the product based on examination over a suitable period of time of a substantial number of lots of the pre-sterilized product.

The D-value (Decimal Reduction Values) is the time (in minutes) required to reduce the microbial population by 90% or 1 log cycle (i.e. to a surviving fraction of 1/10), at a specific temperature. For example, *Bacillus stearothermophilus* is 1.5 minutes under the total process parameters, e.g. 121^0, if it is treated for 12 minutes under the same conditions, it can be said that the lethality input is 8D. The effect of applying this input to the product would depend on the initial microbial burden. Assuming that its resistance to sterilization is equivalent to that of the biological indicator, if the microbial burden of the product in question is 10 micro-organisms, lethality input of 2D yields a microbial burden of 10° (i.e. 1) and a further 6D yields a theoretical survivor probability of 10^{-6}. (Under the same conditions, a lethality input of 12D would amount to an "overkill" approach) In general, the survivor probability achieved for the article under the validated sterilization cycle is not completely correlated with what may occur with the biological indicator. Therefore, for valid use, it is essential that the resistance of the biological indicator be greater than that of the natural microbial burden of the article sterilized.

For any specific sterilization procedure the D-value of the biological indicator should be determined in order to evaluate the resistance of the microbial burdened and to determine the lethality of the process under production conditions. Appropriate sterilization cycles can be established only on the strength of this information.

11.2 Sterilization Methods

The following methods of sterilization are useful for making the sterile packages of pharmacopoeial products. The effect of the chosen method of sterilization on the product (including its final container or package) should be investigated and the procedure validated before being applied in practice. In establishing any process of sterilization that is applied to a product in its final container, it is important to take into account the non-uniformity of the physical and, where relevant, chemical conditions within the sterilizing chamber. The location within the sterilizing chamber that is least accessible to the sterilizing agent must be determined for each loading configuration of each type and size of container or package (such as the coolest part in an autoclave). The minimum lethality delivered by the sterilizing cycle and its reproducibility must also be determined in order to ensure that all loads will consistently receive the specified treatment.

Having conventional a validated procedure, knowledge of its performance in regular use should, wherever possible, be gained by monitoring and suitably recording the physical and where relevant, chemical conditions achieved within the load in the chamber during each sterilizing cycle.

11.2.1 Steam Sterilization (Heating in an Autoclave)

This process involving heating in an autoclave with saturated steam under pressure should be used whenever possible for aqueous preparations and for surgical materials. An aqueous product which is to be sterilized by this method is distributed into suitable containers which are than sealed so as to exclude micro-organisms. It is than exposed to saturated steam for a time sufficient to ensure that the entire contents of each container are maintained for an effective combination of time and temperature to ensure sterility.

The subsequent combinations of temperature and time are normally in use.

Holding temperature (°C)	Minimum holding time (minutes)
115 to 118	30
121 to 124	15
126 to 129	10
134 to 138	3

Other combinations of time and temperature may be used provided that it has been satisfactorily demonstrated that the process chosen delivers an adequate level of lethality when operated routinely within the established tolerances. In establishing the sterilization cycle parameters the F_0 concept may be used.

The F_0 value of a saturated steam sterilization process is the lethality expressed in terms of the equivalent time in minutes at a temperature of 121° delivered by that process to the product in its final containers with reference to micro-organisms processing a Z-value (the change in temperature required to alter the D-value by a factor of 10) of 10. In general for aqueous preparations a microbiologically – validated steam sterilization process that delivers, in total (i.e. including the heating up and cooling down phases of the cycle), an F_0 value of not less than 8 to every container in the load is considered satisfactory. If a product is especially heat-sensitive or on certain circumstances where a process that delivers, in total, the F_0 less than 8 then it is necessary to take great to ensure that an adequate assurance of sterility is consistently achieved. It is essential not only to validate the process microbiologically but also to perform constant, rigorous, microbiological monitoring during routine production to demonstrate that the microbiological parameters are within the established tolerances so as to give a theoretical level of not more than one living micro-organisms per 10^6 containers in the final products.

When surgical materials are sterilized in an autoclave the steam used should neither be superheated nor contain more than 5% of entrained moisture. Appropriate steps should be taken to ensure adequate removal of air. Most dressings are conveniently sterilized by maintaining at a temperature of 134° to 138° for 3 minutes but other suitable combinations of temperature and time may be used.

11.2.2 Dry Heat Sterilization

This process is suitable for heat-stable, non-aqueous products and powders. A preparation to be sterilized by dry heat is distributed in the final containers which are then either finally sealed or temporarily closed to exclude micro-organisms and then heated so as to ensure that the entire contents of each container are maintained for an effective combination of time and temperature to provide an adequate assurance of sterility. Cycles of a minimum of 180° for not less than 30 minutes, a minimum of 170° for not less than 1 hour or a minimum of 160° for not less than 2 hours may be employed but other combinations of time and temperature may be necessary for certain preparations. Containers that have been temporarily closed during the sterilization procedure are then finally sealed using aseptic technique so as to exclude micro-organisms.

Dry heat sterilization is usually carried out in a sterilizing oven designed specifically for the purpose. The oven is supplied with heated, filtered air, distributed uniformly throughout the chamber by convection or radiation and employing a blower system with devices for sensing, monitoring, and controlling the critical parameters. The validation of a dry heat sterilization facility is carried out in a manner similar to that for a steam sterilizer described earlier. Where the unit is employed for sterilising components such as containers intended for intravenous solutions, care should be taken to avoid accumulation of particulate matter in the chamber.

A microbial survival probability of 10^{-12} is considered achievable for heat- stable articles or components.

11.2.3 Gas sterilization

This method involves exposure to a sterilizing gas. The gas generally employed in gaseous sterilization is ethylene oxide (EtO) of acceptable sterilizing quality. The process is difficult to control, the gas itself being toxic and, when mixed with certain proportions with air, explosive. Care should be taken to ensure that the process is applied only to packaging materials compatible with the sterilizing gas.

The sterilizing efficiency of ethylene oxide is related to the concentration of gas, the humidity, and the time of exposure, the temperature and the nature of the load. Knowledge of the performance of the process in routine use should, wherever possible, be gained by monitoring and suitably recording the temperature, the relative humidity and the concentration of ethylene oxide achieved within the load in the sterilizer. The effectiveness of each sterilization cycle should be demonstrated by the use of suitable biological indicators distributed throughout the load.

The packaging material to be sterilized is exposed to ethylene oxide or to a mixture of ethylene oxide with a suitable inert gas. After sterilization, adequate time should be allowed for the dispersal of residual ethylene oxide and other volatile residues. This should be confirmed by determining the residual ethylene oxide, ethylene glycol and ethylene chlorohydrin by a suitable method such as gas chromatography.

One of the principal limitations of the ethylene oxide sterilization process is the limited ability of the gas to diffuse to the innermost product areas that require sterilization. Package design and chamber loading configurations must be determined so that there is negligible resistance to the diffusion of gas.

11.2.4 Sterilization by Ionizing Radiation

Sterilization of drug substances, dosage forms, packaging materials and medical devices may be achieved by exposure in the final containers or packages to ionizing radiation in the form of gamma radiation from a suitable radio-isotopic source such as cobalt-60 or cesium-137 or of electrons energized by a suitable electron accelerator. In either case the radiation dose to yield the required degree of sterility assurance should be established such that within the range of minimum and maximum doses set, the properties of the article being sterilized are acceptable.

For gamma radiation sterilization, an effective sterilizing dose which is tolerated without damaging effect is about 25 kGy (2.5 Mrads). If doses less than 25 kGy are used, additional microbiological monitoring of the product before irradiation will be necessary. In order to validate the efficacy, particularly of the lower exposure levels, it is necessary to determine the magnitude (number and/or degree) of the natural radiation resistance of the microbial population of the product. Specific product loading patterns must be established and absorbed minimum and maximum dosage distribution must be determined.

The radiation dose absorbed by the material being irradiated is best measured by means of suitable chemical or physical methods that are appropriate to the dose-rate range used. Established dosimetry procedures should be employed. These may involve the measurement of induced changes in the optical properties of plastics. Clear plastics or plastics containing radiation sensitive dyes that show colour intensification based directly on the amount of absorbed radiation energy, are suitable but they require careful calibration against a standard source on receipt from the supplier and at appropriate intervals thereafter.

11.3 Factors Affecting Selection of Sterilization Method

Packaging plays an important role in many industries, but few applications areas demanding as those found in the medical device and diagnostic industry. Packaging for medical devices may lack the glamour and glitz of that for consumer goods; however, nowhere else is package integrity as important. The job of the package is to maintain the sterility of the product through its intended shelf life, as well as ensure its efficacy at the time of use. The impressive range of flexible packaging materials available today helps to achieve these goals. This article will discuss tried-and-true methods of packaging, as well as introduce several new technologies for meeting packaging requirements.

There are many important factors to be considered when a packaging system is required. All of these factors will, to a varying extent, influence the choice of materials. Ideally, they should be communicated to the packaging supplier as early as possible in the project.

The first factor to be considered with medical packaging materials is the sterilization method. Many times the packager is committed to a certain method because of product restrictions, process economics, or equipment availability. If end-users are locked into autoclave sterilization, then heat-resistant materials must be used. Some medical products are incompatible with specific sterilization methods. For example, one would not sterilize liquids (such as filled IV bags) with ethylene oxide (EtO), as this process requires permeable packaging materials. For products that are terminally sterilized, the packager's choices are often significantly narrowed based on the sterilization method that will be used.

As stated above, the sterilization method is a vital factor to consider in specifying packaging for medical devices, because once the sterilization method is known, the range of appropriate materials narrows. For each of the major sterilization processes, there are 2-D and 3-D packaging systems in place.

11.3.1 Packaging material and EtO Sterilization

For effective EtO sterilization the packaging material must be breathable to allow the high-humidity EtO gas mixture to infiltrate the package. A partial vacuum is drawn before and after the cycle to facilitate the movement (in and out) of the EtO and moisture vapor. If the package docs not have sufficient permeability, the process will be ineffective. As stresses in the seal area are introduced due to the pressure difference between the inside and outside of the package, a seal failure phenomenon known as sterilizer creep may occur. The package must also withstand the moderately elevated process temperatures, although this is typically not a problem for most materials. A typical 2-D breathable pouch consists of a PET/PE bottom web combined with a Tyvek or paper top web. The Tyvek or paper top web is the breathable portion, and it may be coated or uncoated. Uncoated Tyvek has a Gurley Hill porosity of about 20–25 sec/100 cm3 air per square inch; typical coated Tyvek has a porosity of 80–100 sec/100 cm3 air per square inch. A less-breathable coating can be applied as a patterned coating, such as a grid or dot pattern. Tyvek or paper also provides a suitable print surface and good aesthetics. Because one disadvantage to Tyvek is its expense, a great many 2-D pouch applications use a Tyvek header strip or vent, to save on material cost. For 3-D packages used with EtO sterilization, the same breath ability requirement exists as for 2-D pouches; therefore, the same breathable materials (paper or Tyvek) can be used. Instead of a flat, nonformable bottom web, some type of thermoformable web should be used. The traditional thermoforming webs of EVA/ionomer/EVA are being challenged by more cost effective approaches, such as using linear low-density polyethylene (LLDPE) or Surlyn blends in the core, or even moving to a monolayer blend. The more costly co extruded-nylon forming webs provide outstanding toughness and impact resistance that is not easily matched by using strictly PE-based polymers and blends. To further improve pinhole and puncture resistance, multiple layers of nylon are often used. Given the wide cost range, it is important not to over design the package for the product.

11.3.2 Materials and Steam under Pressure (Breathable Autoclave)

Packaging material requirements for steam sterilizing are similar to those for EtO in that the package must be breathable to allow for steam infiltration. An additional requirement is heat resistance. A typical steam autoclave cycle may be run at 121°C, but temperatures up to 140°C (284°F) may be used. For the breathable portion of the package, Tyvek—which is made of spunbonded HDPE fibers—may be used, but only under very controlled conditions where temperatures do not exceed 127°C. Many papers will withstand the process without significant loss of physical properties. A suitable heat-seal coating could be based on PP or amorphous polyester. In-hospital autoclave sterilizers have long used steam sterilization for 2-D packages. A typical structure is an OPET/PP bottom web and a paper top web. This type of package is very low cost and may not have all the desired properties for in-hospital use such as fiber-free opening and microbial barrier.

Other, more-reliable 2-D pouches employ OPET with a peelable sealant of sufficient heat resistance, based on PP or PET chemistry, and a high-strength, medical-grade paper. A 3-D steam-sterilisable package must be breathable and heat resistant, as well as formable. This is a tall order for most packaging materials, particularly if one desires a peelable seal as well. Nylon-, PP-, or PET-based forming webs may be used with coated paper or Tyvek top webs. Alternatively, uncoated paper or Tyvek can be used if a peelable seal layer is provided as the inner layer on the bottom forming web (a more economical scenario). If the peel layer is part of the forming web, one should be aware that the heat from the forming process can change the nature of this layer. In the case of a peelable polyester sealant, the heat from the forming station can crystallize the sealant, rendering it nonfunctional. There are also the aesthetic issues of shrinkage or snapback with some PP and nylon forming webs when subjected to autoclave temperatures. Developing a more cost-effective, peelable package for 3-D steam sterilization remains an active development area for several converters.

11.3.3 Packaging and Radiation Sterilization

Flexible packaging for radiation sterilization has no requirement for heat resistance (other than for the heat needed to make the seals) or breath ability. Although breath ability is not required to achieve sterility, it is sometimes needed to allow out gassing of odors that can develop. In these cases, a Tyvek or paper top web can be used. The main limitation is that some polymers are not suited for exposure to radiation. Materials that may be adversely affected include PP and polyvinyl chloride (PVC). Adverse effects, which can occur to a modest extent in many materials, usually involve a loss of physical properties (such as embitterment) or a color change. Packaging for radiation sterilization can be supplied in high barrier or no barrier versions. A high-barrier package would be used where the packaged product needs to be protected from oxygen, moisture vapor, and/or light. For a barrier package, both top and bottom webs will utilize a barrier material. For 2-D no barrier radiation pouches, the packaging materials can be very cost-effective. A simple PET/PE laminate can be used for both sides of the web.

The PE side can be supplied to give either a peelable seal or weld seal. Prefabricated linear-tear or snap-tear bags, made from a single coextruded film, can be used. For barrier applications, one would need to incorporate aluminum foil or one of the other materials mentioned above in the discussion on barrier materials. For 3-D radiation, a formable bottom web is used. For nonbarrier applications, the same webs used as bottom forming webs for 3-D EtO packages are employed. For barrier applications, one of the formable barrier webs, such as a coextruded film containing EVOH, an Aclar lamination, or a formable foil, can be used.

11.3.4 Packaging Materials and Retort Sterilization (No breathable Autoclave)

In some no breathable autoclave applications, a liquid-containing package is sterilized in a retort process. Seal integrity is extremely important in these applications because, unless the engineer has precise overpressure control, the pressure difference between the inside and outside of the package may easily result in seal failures. This is an area where very strong peelable seals are often specified. The same high heat resistance needed for breathable steam applications is also required. In general, a barrier package for this type of sterilization can use either clear barrier or foil barrier composites. Not all of the barrier materials are autoclaveable, and not all of the autoclaveable materials are thermoformable. Table I indicates which barrier materials apply for retort and autoclave sterilization. A 2-D barrier pouch for retort or autoclave sterilization can be made from aluminum foil or from one of the nonfoil autoclaveable barrier materials. Autoclaveable clear barrier materials are often more expensive than foil; therefore, a pouch can be made of foil on one side and the clear barrier material on the other. This will suffice provided the end-user does not need clarity on both sides of the package and the packaging equipment can accommodate two different webs. For this sterilization method, 3-D packaging usually consists of a rigid thermoformed tray and a flexible film lidstock. The lidstock can be foil based or use a nonfoil barrier material.

11.3.5 Packing Materials and Other Sterilization Methods

Although the previously mentioned sterilization methods represent the major applications, other methods are well worth mentioning. Dry-heat sterilization is used for products that do not lend themselves well to other sterilization techniques, do not contain moisture, may be damaged by contact with moisture, and are stable at elevated temperatures. This technique requires a long exposure time (as long as several hours) at elevated temperatures (275°F and above). Heat resistance is the main packaging requirement. Some examples of dry-heat-sterilized products include orthopedic implants, collagen products, and surgical instruments. Hydrogen peroxide (H_2O_2) gas plasma sterilization is drawing interest as a substitute for EtO sterilization. Developed by Advanced Sterilization Products (Irvine, CA), the Sterrad system is gaining acceptance in both institutional and industrial use. A breathable package is required for the gas to permeate the package. Although this is done in a drier and slightly cooler environment than EtO sterilization, the same types of package materials are appropriate. The only known exception is paper, which cannot be used; the cellulose fiber in paper can absorb enough H_2O_2 to make the process ineffective. Since this method is relatively new, not all materials have been thoroughly tested. One of the main benefits of this process is greatly improved cycle time; since the primary by-products are water and oxygen, there is no aeration required. This is very important to help achieve quick turnaround in clinical settings. Compared to EtO, the environmental and regulatory concerns are minimal. Pure Pulse Technologies (San Diego) has developed Pure- Bright broad-spectrum pulsed light (BSPL) technology, a sterilization process used primarily in applications involving blood components, vaccines, and other biopharmaceuticals, as well as dry surfaces. Highly intense, short flashes of broad-spectrum light (ultraviolet, visible, and infrared) are used to rapidly deactivate organisms without the use of intense heat, hazardous chemicals, or by-product generation. Packaging materials compatible with this process will transmit light over the broad spectrum employed. LLDPE, LDPE, nylon, Aclar, HDPE, and PP have all been used.

Polystyrene, PET, and glass are all very clear to the eye, and one might expect these materials to work well. In reality, these materials do not transmit light in the ultraviolet region of the spectrum and, therefore, are not suitable for use with this method. Package geometry should not allow any shadowing on the product or the light exposure may not be sufficient.

CHAPTER 12

Packing of Parenterals, Ophthalmic and Aerosols

Summary

12.1 Introduction

The FDA's requirements, as spelled out in the "Guidance Container Closure Systems for Packaging Human Drugs and Biologics," include levels of extractables/leachables and test methods related to these contaminants, raising the bar on what is expected from biopharmaceutical drug sponsors.

12.1.1 Packaging and Product

Many biopharmaceuticals are sensitive to silicone oil, a material commonly used to lubricate elastomeric stoppers during fill/finish to facilitate insertion of the stopper into the vial. Silicone oil has been associated with protein inactivation through nucleation of proteins around oil droplets

12.1.2 Sources of Contamination

Package component fabricators evaluate extractables from their materials as part of their development and qualification operations. More importantly, leachables tests are carried out at the point of use, in real-life situations in the presence of the actual drug product.

12.1.3 Mitigating the Risk from Rubber Closures

Fluorocarbon film coatings provide the best combination of protection from extractables from the stopper material while providing a high level of barrier protection for the drug product, therefore minimizing leachables.

12.1.4 Strategies for Minimizing Risk

Drug developers who do not understand the impact of packaging on their biopharmaceutical products are courting an unnecessary level of regulatory and product-related risk. Problems often arise in this regard when a contract manufacturer tries to convince a sponsor that a particular stopper, vial, or other closure product is appropriate because it has been validated with the contractor's fill line.

12.1.5 Lyophilization a Special Case

Many biotech products are lyophilized in the package before the stopper and seal are introduced. Lyophilization presents its own peculiar process and packaging requirements.

12.2 Types of Packages

For packaging of parenteral products the types of packages generally are as follows

1. Vials and Bottles
2. Ampoules
3. Prefilled Syringes

12.3 Packaging of Aerosols

The aerosol containers are made from glass and drawn Aluminium. Usually for pharmaceutical product is made up of drawn Aluminium. A hydrocarbon is used as a propellant. If there are compatibility problem between the product and the container, the inner side of the container may be coated. On the upper top size of the container is fixed a spray nozzle, attached to a dip tube, which touches the product. The spray nozzle is usually metered to allow a specific dose to be dispensed with each spray. These are used for inhalers, etc.

12.4 Packaging Procedure for Aerosols

A. Pressure Filling
B. Cold Filling
C. Under cap filling

12.1 Packaging of Parenterals and Ophthalmic

The unit dose for many injectable biotech products is the single-dose vial, with prefilled syringes a distant second. Product is provided either as a solution or, more commonly, as a lyophilized cake, which the caregiver reconstitutes and injects via syringe.

Requirements for product purity, activity, and shelf life dictate a high standard for injectable drug packaging, particularly for highly active peptides and proteins. Packaging represents the first line of defense for all formulated pharmaceuticals, protecting the product from the outside world and vice versa. At the same time, the package must be fully compatible with the product, whether it is in solution or lyophilized.

The FDA's requirements, as spelled out in the "Guidance Container Closure Systems for Packaging Human Drugs and Biologics," include levels of extractables/leachables and test methods related to these contaminants, raising the bar on what is expected from biopharmaceutical drug sponsors.

To protect and ship highly sensitive pharmaceutical powders and liquids, drug manufacturers rely on high-barrier vials, bottles, and syringes. These packages must keep the pharmaceuticals sterile yet allow healthcare practitioners to use them efficiently, safely, and quickly. To ensure that such packages remain sterile, both industry groups and FDA are working on guidance that assist both the manufacturers and the users of parenteral and ophthalmic packaging.

Industry is also involved by developing high-tech aseptic systems such as barrier isolators to nearly eliminate contamination associated with the filling and sealing of vials, bottles, and syringes. Industry is also creating systems that help users deliver drugs quickly and efficiently yet minimize accidents such as needle sticks and overdoses.

The high value, clinical efficacy, and price tags for biopharmaceuticals, coupled with injectable delivery in most cases, demand a high level of awareness of primary packaging. Biotech companies entering the clinical stage need to take the same science- and risk-based approach to packaging materials as they exercise with molecule development. Where that expertise is lacking in-house, developers of biotherapeutics must look outside their organizations for the know-how and experience to assure smooth transition from lab to clinic to market.

Specifying advanced coatings, for example fluoroelastomers, for most stoppers or plungers used with lyophilized or solution-based therapeutic proteins and peptides may seem like an extravagance. In reality, given the long development times and consequences of being wrong, these measures are actually prudent and will lower costs in the long run.

With product safety in mind, both industry and FDA are working to improve parenterals and ophthalmic packaging. Designed to protect latex-sensitive patients, a new FDA ruling regarding the labelling of latex-containing packages affects pharmaceutical manufacturers, who will need to either institute the new labelling or find nonlatex alternatives. New guidelines from FDA and PDA along with developments in barrier isolation technology should help industry refine filling and packaging processes. And industry is devising safer, more reliable ways to package and deliver drugs. As much as these developments affect pharmaceutical packagers, their impact will be felt most by healthcare practitioners, who will be able to deliver drugs more safely and efficiently while minimizing nosocomial infections and allergic reactions.

12.1.1 Packaging and Product

Modern biopharmaceuticals are overwhelmingly proteins and peptides, molecules with unique chemical, physical, and mechanical properties. Proteins are sensitive to heat, light, and chemical contaminants. Minute concentrations of metals, plasticizers, and other materials from packaging may deactivate or denature therapeutic proteins. Proteins and peptides have a tendency to adsorb onto the surface of packaging containers and closures, which can essentially remove all active material from the drug formulation. In situations where the drug desorbs back into the solution, the interaction could cause the drug to lose potency.

Lyophilized proteins are no less immune from the effect of packaging. Since most lyophilization cakes are sensitive to moisture, an inadequate seal could cause water and other contaminants to enter the package and deactivate the drug.

Many biopharmaceuticals are sensitive to silicone oil, a material commonly used to lubricate elastomeric stoppers during fill/finish to facilitate insertion of the stopper into the vial. Silicone oil has been associated with protein inactivation through nucleation of proteins around oil droplets. Recently introduced fluoroelastomer coatings on stoppers provide needed lubricity in addition to an added level of chemical inertness, barrier protection, and safety. Fluoroelastomers thus serve as both a lubricant and a barrier to improve compatibility between product and the rubber closure.

12.1.2 Sources of Contamination

Extractable are the most common source of leachables contamination. An extractable is a chemical species, released from a container or component material, which has the potential to contaminate the pharmaceutical product. Extractables are generated by interaction between strong solvents and the package (including the glass vial and stopper) over time depending on temperature and extraction conditions.

A leachable is a chemical that has migrated from packaging or other components into the dosage form under normal conditions of use or during stability studies.

Package component fabricators evaluate extractables from their materials as part of their development and qualification operations. More importantly, leachables tests are carried out at the point of use, in real-life situations in the presence of the actual drug product.

The potential impact of extractables and leachables on drug products is significant, especially with highly active biopharmaceutical drug products, which may contain just femptograms of active ingredient.

12.1.3 Mitigating the Risk from Rubber Closures

Fluorocarbon film coatings provide the best combination of protection from extractables from the stopper material while providing a high level of barrier protection for the drug product, therefore minimizing leachables.

When applied to stoppers, fluorocarbon films reduce adsorption of the drug onto the stopper, which is critical for maintaining the product's potency and shelf life. In addition, fluorocarbon films provide extra lubricity for proper vial seating, without the need for silicone oil. Fluoroelastomer films, which are made from inert materials, also reduce the possibility of extractable migrating from the rubber stopper into the biopharmaceutical product.

Since the cost of specifying the wrong closure components and materials is so high, biopharmaceutical manufacturers need to devise a separate development plan for primary packaging, just as they do for molecule and clinical development.

Normally this separate activity is contracted out to firms that specialize in packaging components by Phase I when sponsors and regulators get serious about product and package. During Phase I a package component expert company will begin screening for closure designs and materials.

Screening involves assessing packaging alternatives, generating preliminary data on leachables, and choosing one or several alternatives that provide the highest degree of product compatibility and the lowest level of leachables. By Phase II, earlier if possible, sponsors need to begin to develop precise, validated methods for determining extractables and leachables. For products that get this far, methods development becomes almost a separate phase of stability testing.

It is difficult to overestimate the importance of carrying out these studies for the full testing period. Some product-package combinations that showed little or no degradation over the first few months may lead to significant inactivity, due to adsorption onto the glass vial prior to expiration of a two-year shelf life. Similarly, leachables that do not appear for the first several weeks may emerge later on, well within the products specified shelf life.

12.1.4 Strategies for Minimizing Risk

Drug developers who do not understand the impact of packaging on their biopharmaceutical products are courting an unnecessary level of regulatory and product-related risk. Problems often arise in this regard when a contract manufacturer tries to convince a sponsor that a particular stopper, vial, or other closure product is appropriate because it has been validated with the contractor's fill line. That is all well and good, and even necessary. However, stoppers need to be validated with the product first, and only then with the filling machinery. It is far more prudent, and in the long term much more cost-effective, to test and validate packaging within the context of the drug product.

12.1.5 Lyophilization a Special Case

Many biotech products are lyophilized in the package before the stopper and seal are introduced. Lyophilization presents its own peculiar process and packaging requirements.

As with solution-phase biopharmaceuticals, packaging can make or break final formulation for lyophilized products, particularly with respect to the product's long-term stability and compatibility with the package. Vials that are not designed specifically for lyophilization, for example with convex rather than flat bottoms, make the lyophilization process less efficient, leading to an extended lyophilization cycle. Rubber closures can also hinder freeze-drying if they do not permit adequate venting during sublimation.

Stopper rubbers adsorb and desorb water at different rates. Under storage conditions stoppers that were not properly dehydrated can release water into the lyophilized product, affecting product stability over time. This can be especially problematic with lyophilized biopharmaceuticals, which tend to have small cake weights when compared to traditional pharmaceuticals following lyophilization. Since their weight is often in the range of milligrams or less, these cakes are significantly more sensitive to moisture, pH changes, and extractables that migrate from the rubber closure.

A small difference in moisture in the lyophilization cake can make the difference between an active and denatured protein. pH differences, which may be caused by contaminants, can seriously affect protein structure and activity. The wrong rubber closure can easily shift pH units in a small volume of product or a diluted lyophilization cake.

Fluoroelastomer-coated stoppers eliminate the rubber closure as a source of the leachable that could impact pH because of its barrier properties. Glass vials, however, can also leach ions, which can impact pH.

Whatever precautions are taken with solution-phase preparations are doubly applicable to lyophilized biopharmaceuticals. During lyophilization all the primary package components must work together without interfering with either the product or the process. Some packaging issues to be aware of for lyophilized products include closures that allow adequate sublimation rates and cleanly insert into the vial without "back out" or sticking to the lyophilization chamber shelves; glass vials that provide adequate contact between the base of the vial and the lyophilization shelf; and compatibility during lyophilization between vial and elastomeric closure.

12.2 Types of Packages

For packaging of parenteral products the types of packages generally are as follows

1. Vials and Bottles
2. Ampoules
3. Prefilled Syringes

1. Vials and Bottles

Vials and Bottles are the containers in which the multiple dose parenteral preparations are packed. The material of these containers is generally glass. The vials are provided with a closure followed by an aluminum seal to ensure perfect air tight packing. Some examples rials and bottles available in market are shown in Fig. 12.1 to Fig. 12.4.

Fig. 12.1 Filled Vials.

Fig. 12.2 Vials with rubber closure.

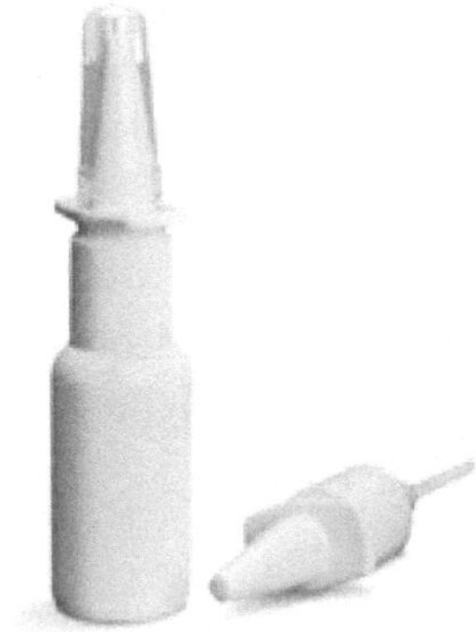

Fig. 12.3 Nasal and Ophthalmic Packing.

Fig. 12.4 Bottles for ophthalmic Drugs.

2. Ampoules

A single dose parenteral preparation is packed in an ampoule type of container, generally made up of glass. (Fig. 12.5). Recently plastic ampoules for 'water for injection' are available in market (Fig. 12.6).

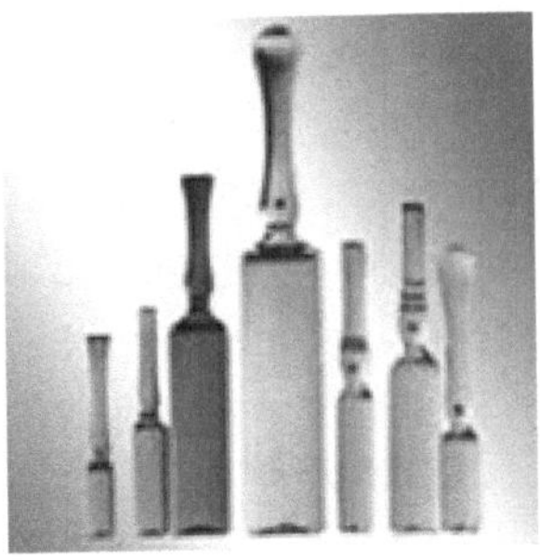

Fig. 12.5 Filled and Sealed Ampoules.

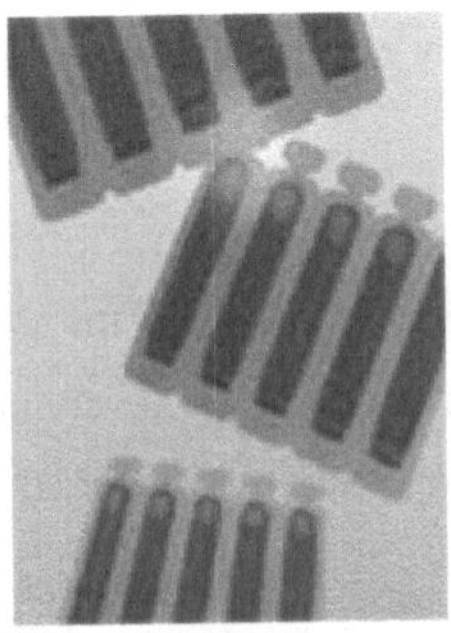

Fig. 12.6 Plastic Packing of Parenterals.

3. Prefilled Syringes

This type of packaging system is usually employed on small dose parenteral preparations. The drug is already filled in the syringe and it is ready for administration. (Fig. 12.7). Generally diagnostic agents and vaccines are packed in this type of packages. The materials for syringes may be glass or plastic.

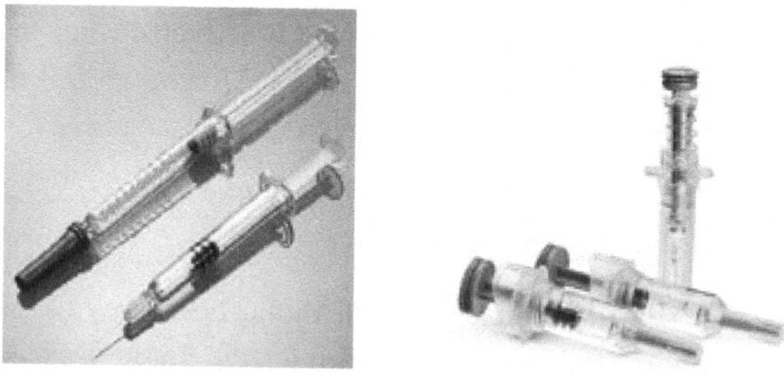

Fig.12.7 Prefilled Syringes.

12.3 Packaging of Aerosols

The aerosol containers are made from glass and drawn Aluminium. Usually for pharmaceutical product is made up of drawn Aluminium. A hydrocarbon is used as a propellant. If there are compatibility problem between the product and the container, the inner side of the container may be coated. On the upper top size of the container is fixed a spray nozzle, attached to a dip tube, which touches the product. The spray nozzle is usually metered to allow a specific dose to be dispensed with each spray. These are used for inhalers, etc. An aerosol container with spray nozzle is shown in Fig. 12.8.

The aerosols can be formulated in various dispensing systems, and accordingly package has to be modified. The common aerosol formulation systems are solution system (Homogeneous), suspension or dispersion system, foam system etc. The diagrammatic representations of some of these packaging design systems are illustrated in Fig. 12.9, 12.10(a), 12.10(b) and Fig. 12.11.

12.3.1 Bottle Seals

Bottle seals, which are tamper resistant, use an inner seal that a bonded to the neck of the bottle and will get destroyed once the bottle is opened.

Several variants of such inner and tamper-proof seals have been developed over the time. Usually a lamination, which is twin layered, is bonded together with an adhesive. Such a lamination is inserted over the bottle cap using wax or the rim of the bottle. The application of the cap forces the inner seal into contact with the glued rim the bottle. Inner seals can be pressure sensitive also and during capping the adhesive bonds inner seal to the rim of the bottle. Apart from this heat sensitive adhesives can also be used for the purpose of tamer-proofing which is activated when the bottle is passed under a heating induction coil. Usually Aluminium foil is used for this kind of sealing and plastic cap are preferred over metal caps, which can rupture the foil.

12.3.2 Breakable Caps

Breakable closure are being used for alcoholic beverages like beer etc.; where an Aluminium foil shell is placed over the bottle neck and passes over the bottle closure to the neck of the bottle and is pressure sealed, To remove the closure the bottom potion of the closure will have to be torn off. These are used for solvents, etc. For pharmaceutical purpose roll-on-pilfer-proof (ROPPF) caps are mostly used

12.3.3 Sealed Tubes

Collapsible plastic, metal or laminated tubes are used for packaging. Metal tubes are used where the product is moisture sensitive and needs a high degree of barrier. These are usually made of Aluminium. Inserts are used to seal the tube and make it puncture resistant. If any attempt is made to tamper with such a product the inserts will be punctured and pierced open. Extruded plastic tubes are also used. They are filled from the other end are sealed by crimping the ends in the case of metal tubes or by induction sealing in case of laminated or plastic tubes.

12.3.4 Tape Seals

This type of sealing requires the application of a pressure-sensitive tape over the package closure, which will be ruptured off when the package is opened. High-density light-weight paper is usually used for this purpose as these have poor tear strength and cannot survive any attempt at removal once they are applied.

12.3.5 Container Requirement

Containers must with stand pressure asdhighas140-180 psig at 130^0F

Fig. 12.8 Aerosol pack.

12.3.6 Types of Materials for Aerosol Containers

(a) Tin plated steel
   1. Three piece – soldered side seam
   2. Two piece- welded side seam or drawn
   3. Drawn and Ironed

(b) Aluminium
   1. One piece – Extruded
   2. Two piece – extruded drawing

(c) Stainless steel.

(d) Glass Uncoated

(e) Plastic coated glasses.

(f) Plastic and synthetic resin

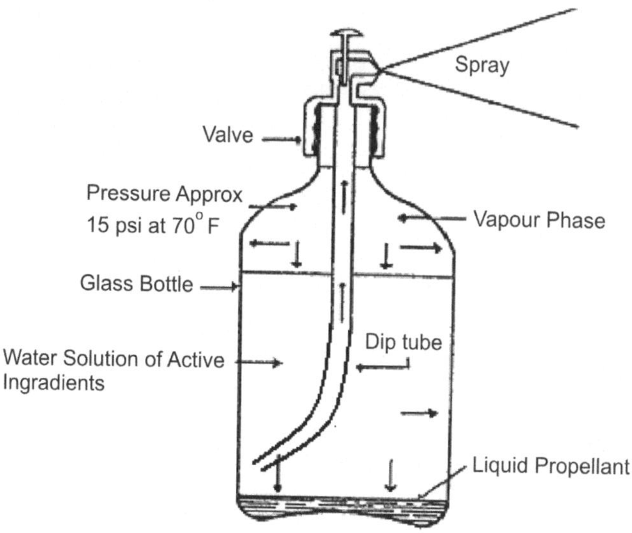

Fig. 12.9 Aerosol packaging: three phase system.

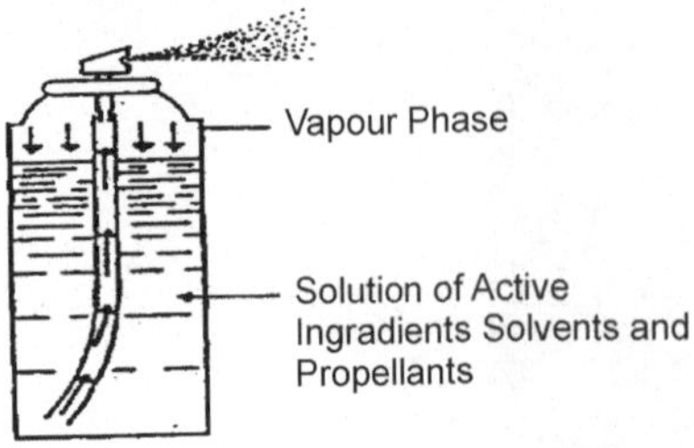

Fig. 12.10 (a) Homogeneous aerosol

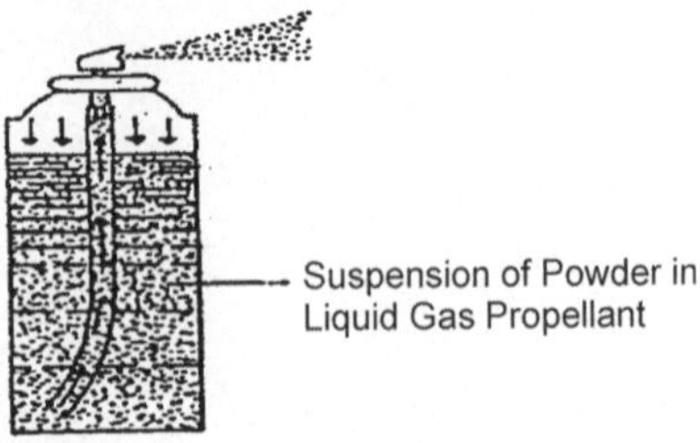

Fig. 12.10 (b) An aerosol powder spray.

12.3.7 Valve Assembly

The valve in the aerosol facilities the dispensing of the product besides ensuring a hermetically seal when the aerosol is not in use. Different type of valve dispensed the product as- sprays, solid streams and foams.

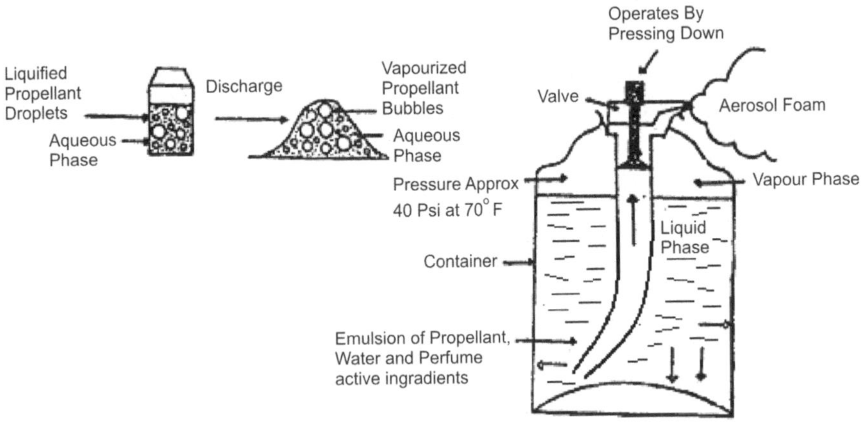

Fig. 12.11 Foam forming aerosol pack.

The principles implicated in the construction of valves for glass, plastic or metal containers are essentially the same. Followings are the various components of an aerosol valve (Fig. 12.12):

- Actuator
- Mounting Cup
- Valve stem (core)
- Valve stem (seat) gasket
- Spring
- Housing (valve body)
- Dip tube

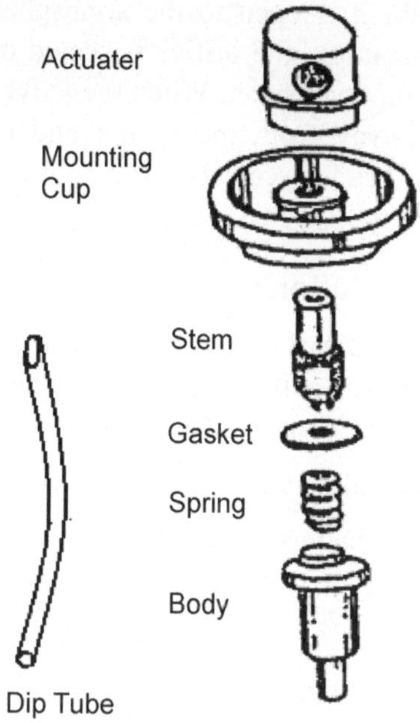

Fig. 12.12 Components of aerosol valve.

12.3.8 Types of Valves

Mainly there are two types of valves are common in packaging of Pharmaceutical packaging, which are further classified into:

1. **Common Valves (Continuous Discharge)**
   - Vertical action, sliding gasket, "Male" valves.
   - Vertical action, deflecting gasket, 'Male valves
   - Tilt action valves.
   - Vertical action, sliding, gasket, "Female" valves
2. **Special Valves**
   - Spray-any-way valves
   - Tip-sealing valves
   - Metering Valves
   - Vapour tap valves
   - Foam valves
   - Powder valves.
   - Compressed gas valves

The valve stem, which is open to the atmosphere, containers a small orifice. In the closed position, the orifice is sealed off from the interior of the aerosol by an elastomeric gasket. When the valve stem is pushed down, the orifice is moved away from the gasket and down into the aerosol container. The aerosol product can then flow through the orifice and out into the atmosphere.

12.3.9 Actuators or Buttons

The actuator is the molded part on top of the valve stem that the consumer pushed open the valve. The actuator is a significant factor in determining whether the product will be dispensed as a foam, spray or stream.

Different types of actuators are

1. Solid stream Actuators
2. Spray Actuators
3. Foam Actuators
4. Metered – Dose Inhalers.

12.3.10 Dip-Tube

Dip-tubes are made from polyethylene or polypropylene, providing it should be inert with the ingredients of the products.

12.3.11 Product Concentrate

The product concentrate consists of active ingredients or a mixture of active ingredients and other necessary agent such as solvents, antioxidants and surfactants.

12.4 Packaging Procedure for Aerosols

Three methods have been used to package aerosol products. All aerosol concentrate are liquids or powders at atmosphere pressure and are loaded directly into the aerosol container before the valve is added.

Propellants are gases at room temperature and atmospheric pressure and are liquefied before loading. There are three different methods.

(a) Pressure Filling Method

Propellants are liquefied by pressure and force into container after the valve has been crimped (Fig. 12.13). It is mint for product that could not be cold filled. These included aqueous-based products that froze, for example-shaving lather, foam analgesics. The following steps are involved

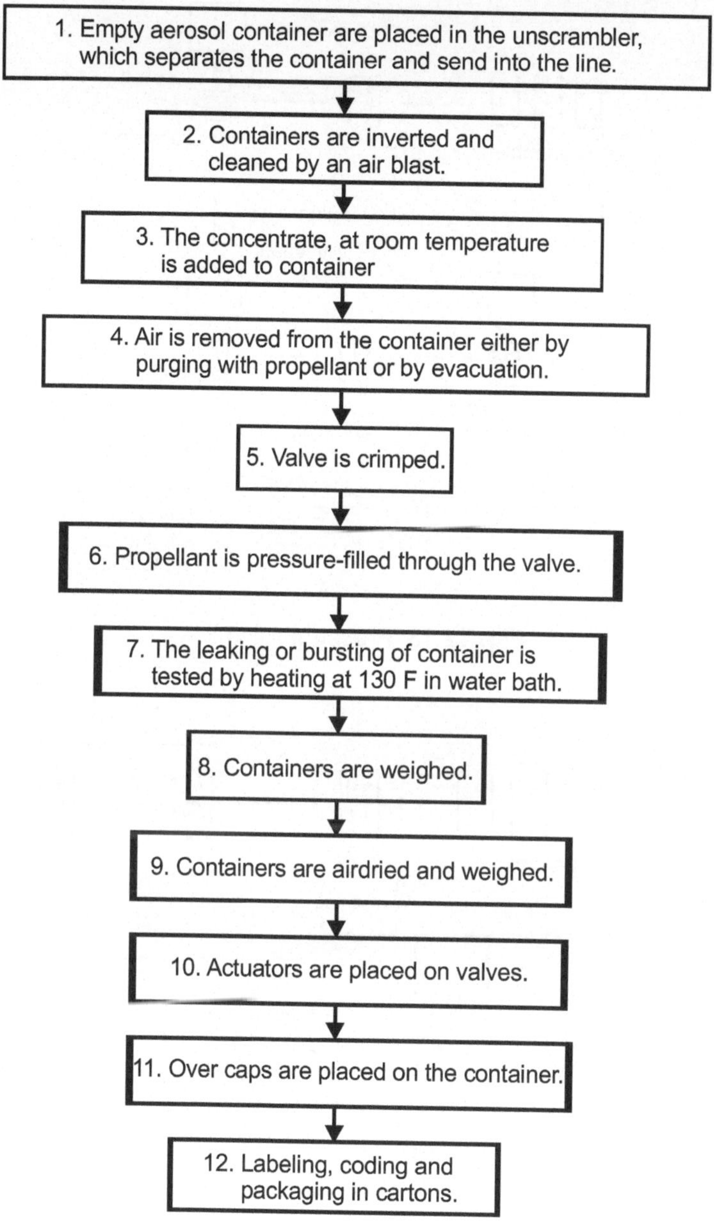
1. Empty aerosol container are placed in the unscrambler, which separates the container and send into the line.
2. Containers are inverted and cleaned by an air blast.
3. The concentrate, at room temperature is added to container
4. Air is removed from the container either by purging with propellant or by evacuation.
5. Valve is crimped.
6. Propellant is pressure-filled through the valve.
7. The leaking or bursting of container is tested by heating at 130 F in water bath.
8. Containers are weighed.
9. Containers are airdried and weighed.
10. Actuators are placed on valves.
11. Over caps are placed on the container.
12. Labeling, coding and packaging in cartons.

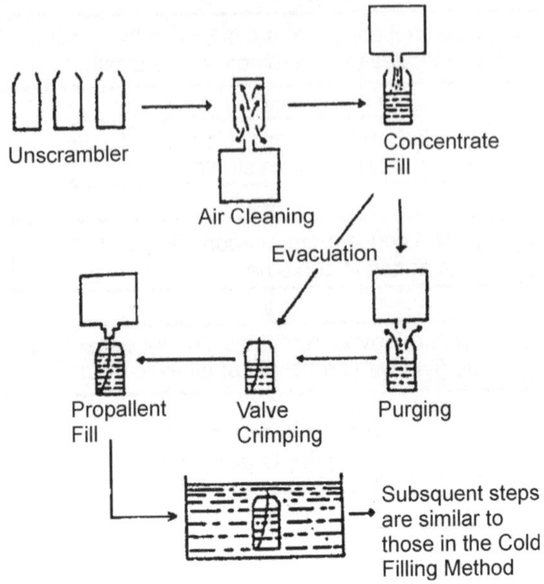

Fig. 12.13 Pressure filling line.

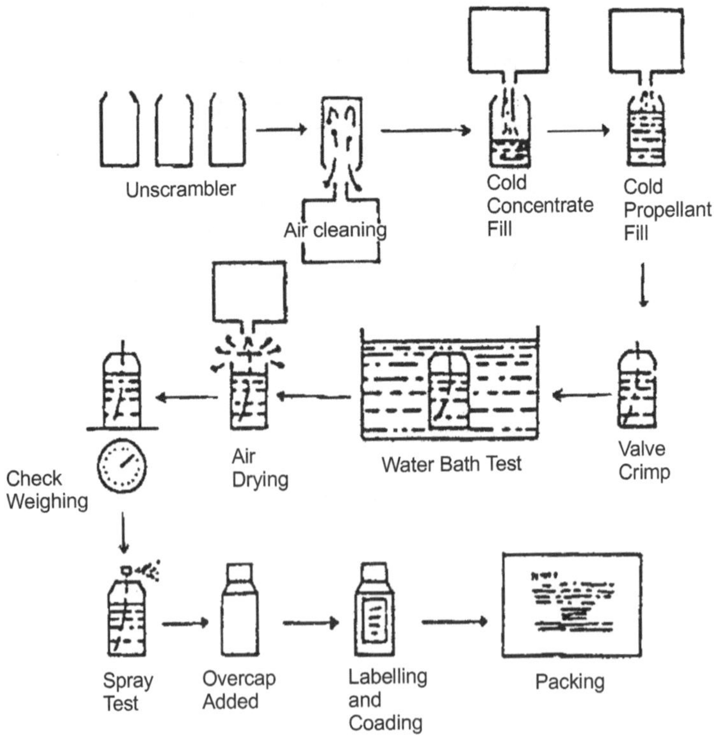

Fig. 12.14 Cold filling line.

(b) Cold Filling

Propellants are liquefied by cooling, added to aerosol container and then valve is crimped (Fig. 12.14). The following steps are involved

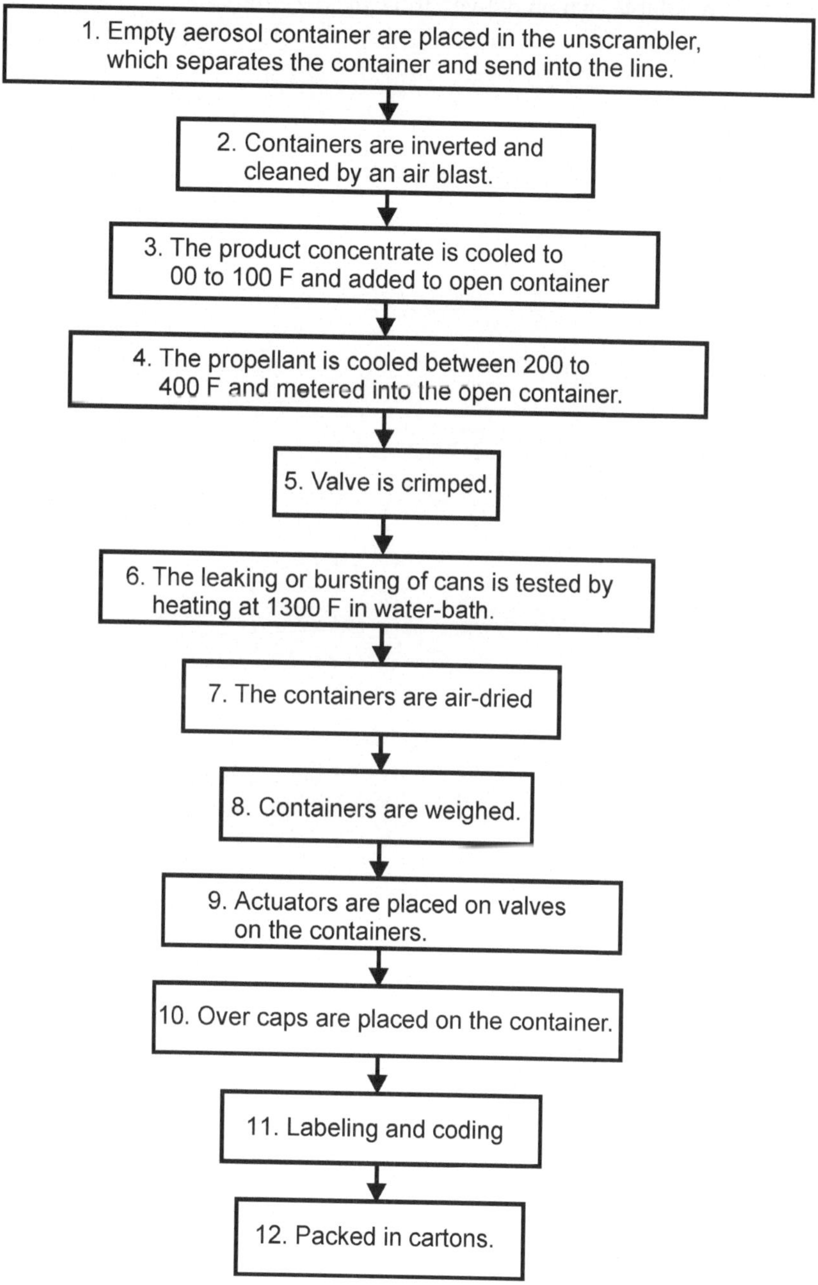

(c) Under Cap Filing

This combines the best elements of both cold and pressure filling method. Rotary machine with multiple under cap filling heads are available, which deliver 300 cans per minute.

CHAPTER 13

Corrugated Fiberboard and Cartons

Summary

13.1 Introduction

In the mid-19th century, an ingenious concept enabled flimsy sheets of paper to be transformed into a rigid, stackable and cushioning form of packaging for delicate goods in transit. Corrugated (also called pleated) paper was patented in England in 1856, and used as a liner for tall hats, but corrugated boxboard would not be patented and used as a shipping material until December 20, 1871. The patent was issued to Albert Jones of New York, New York for single-sided (single-face) corrugated board.

13.2 Corrugated Fiberboard

Corrugated fiberboard is a paper-based construction material consisting of a fluted corrugated sheet and one or two flat linerboards. It is widely used in the manufacture of corrugated boxes and shipping containers.

13.3 Manufacture of Corrugated Board

Corrugated board is manufactured on large high-precision machinery lines called Corrugators running at 500 lineal feet per minute or faster.

The corrugated medium is usually a 26 lb/1000 sq ft (127 g/m^2) paperboard; higher grades are also available. It arrives to the corrugator on large rolls. At the single-facer, it is heated, moistened, and formed into a fluted pattern on geared wheels.

- **Box Manufacture:** Boxes can be formed in the same plant as the corrugator. Alternatively, sheets of corrugated board may be sent to a different manufacturing facility for box fabrication

- **Single Phase Laminate:** Specialized equipment is necessary for the material construction of SFL, so users may expect to pay a premium for these products.
- **Recycling:** Old corrugated containers are an excellent source of fiber for recycling. They can be compressed and baled for cost effective transport to anywhere in need of fiber for papermaking.

13.4 Carton

As these were extremely valuable, often commanded by the very richest art-buyers, including princes who hung them in their palaces and even took them on their travels as prestigious displays of wealth, often with a visual message, especially the world-famous Flemish ateliers were deemed worthy to have cartons made by some of the greatest graphic artists of the time, including such celebrated painters as Rubens.

13.5 Paperboard

Paperboard **is a paper**-like material, usually over ten mils (0.010 inch) in thickness. Some types of paperboard (corrugated medium and linerboard) are used in the construction of corrugated fiberboard. Other types of paperboard are used for folding cartons, set-up boxes, carded packaging, etc.

Cardboard is a generic, non-specific, lay term used to refer to any heavy paper-pulp based board, such as card stock, or to corrugated fiberboard.

13.1 Introduction

Corrugated fiberboards are produced continuously at high speed by a corrugating machine on which a first linerboard is bonded to the top at one side of flutes of a corrugated medium to form a single faced corrugated fiberboard to which a second linerboard is bonded to the top at the other side of flutes of corrugated medium.

In the mid-19th century, an ingenious concept enabled flimsy sheets of paper to be transformed into a rigid, stackable and cushioning form of packaging for delicate goods in transit.

Corrugated (also called pleated) paper was patented in England in 1856, and used as a liner for tall hats, but corrugated boxboard would not be patented and used as a shipping material until December 20, 1871. The patent was issued to Albert Jones of New York, New York for single-sided (single-face) corrugated board. Jones used the corrugated board for wrapping bottles and glass lantern chimneys. The first machine for producing large quantities of corrugated board was built in 1874 by G. Smyth, and in the same year Oliver Long improved upon Jones' design by inventing corrugated board with liner sheets on both sides. This was now corrugated board as we know it today.

The Scottish-born Robert Gair invented the corrugated box in 1890; that is, pre-cut flat pieces manufactured in bulk that folded into boxes. Gair's invention, as with so many other great innovations, came about as a result of an accident: he was a Brooklyn printer and paper-bag maker during the 1870s, and one day, while he was printing an order of seed bags, a metal ruler normally used to crease bags shifted in position and cut them. Gair discovered that by cutting and creasing bags in one operation he could make prefabricated paper boxes. Applying this idea to corrugated boxboard was a straightforward development when the material became available. By the start of the 20th century, corrugated boxes began replacing the custom-made wooden crates and boxes previously used for trade.

The corrugated box was initially used for packaging glass and pottery containers, which are easily broken in transit. Later, the case enabled fruit and produce to be brought from the farm to the retailer without bruising, improving the return to the producers and opening up hitherto unaffordable export markets. (There had previously been a great deal of waste when, for example, oranges were craned out of the hold of a ship, having been bulk loaded into it.)

13.2 Corrugated Fiberboard

Corrugated fiberboard is a paper-based construction material consisting of a fluted corrugated sheet and one or two flat linerboards. It is widely used in the manufacture of corrugated boxes and shipping containers.

Fig. 13.1 Corrugated fiberboard.

The corrugated medium and linerboard are made of paperboard, a paper-like material usually over ten mils (0.010 inch, or 0.25 mm) thick. Paperboard and corrugated fiberboard are sometimes called *cardboard* by non-specialists; although cardboard might be any heavy paper-pulp based board. (Fig. 13.1).

13.3 Manufacture of Corrugated Board

Corrugated board is manufactured on large high-precision machinery lines called Corrugators running at 500 lineal feet per minute or faster.

The corrugated medium is usually a 26 lb/1000 sq ft (127 g/m^2) paperboard; higher grades are also available. It arrives to the corrugator on large rolls. At the single-facer, it is heated, moistened, and formed into a fluted pattern on geared wheels. This is joined to a flat linerboard with a starch based adhesive to form single face board. At the double-backer, a second flat linerboard is adhered to the other side of the fluted medium to form single wall corrugated board. Linerboards are often Kraft paperboard (of various grades) but may be bleached white, mottled white, colored, or preprinted.

Common flute sizes are "A", "B", "C", "E" and "F" or microflute. The letter designation relates to the order that the flutes were invented, not the relative sizes. Flute size refers to the number of flutes per lineal foot. For example, "B" flute is approximately 1/4 inch from the top of one flute to the next, or 50 flutes per foot. "C" Flute is 5/16 inch from flute to flute or 42 flutes per lineal foot. "E" flute is 1/8 inch flute to flute or 90 flutes per lineal foot. Board thickness is an unreliable metric, due to various manufacturing conditions. However, a rough guide is: "C" flute=5/32 inch thick, "B" flute=1/8 inch thick, "E" flute=1/16 inch thick. The most common flute size in corrugated boxes is "C" flute.

Corrugated board is often graded by the basis weights of the linerboards, burst or mullen strength, edge crush test, or flat crush test. TAPPI and ASTM test methods for these are standardized.

The choice of corrugated medium, flute size, combining adhesive, and linerboards can be varied to engineer a corrugated board with specific properties to match a wide variety of potential uses. Double and triple-wall corrugated board is also produced for high stacking strength and puncture resistance.

Box Manufacture

Boxes can be formed in the same plant as the corrugator. Alternatively, sheets of corrugated board may be sent to a different manufacturing facility for box fabrication.

The corrugated board is creased or scored to provide controlled bending of the board. Most often, slots are cut to provide flaps on the box. Scoring and slotting can also be accomplished by die-cutting.

The "Flexo Folder Gluer" is a machine that in one single pass prints, cuts, folds, and glues flat sheets of board to convert them to boxes for any application, from storing old family pictures to shipping the biggest of plasma TV sets to the global market. The most advanced of FFG's can run at speeds of up to 26,000 boxes per hour.

The most common box style is the Regular Slotted Container. All flaps are the same length and the major flaps meet in the center of the box.

Box blank slotted container

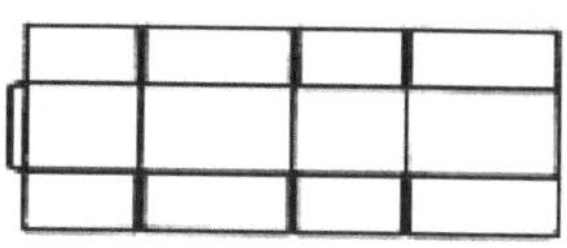

Fig. 13.2 Box blank showing score lines, slots, and manufacturer's joint.

The manufacturer's joint is most often joined with adhesive but may also be taped or stitched. The box is shipped flat (knocked down) to the packager (Fig. 13.2) who sets up the box, fills it, and closes it for shipment (Fig. 13.3). Box closure may be by tape, adhesive, staples, strapping, etc.

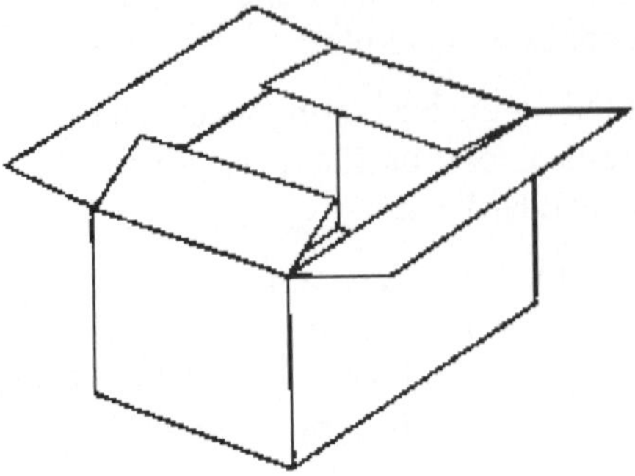

Fig. 13.3 Corrugated fiberboard box.

Many other styles of corrugated boxes and structures are available

- A Full Overlap Box is similar to an RSC except the major flaps fully overlap.
- A Five Panel Folder is shipped flat without a manufacturer's joint and is often used for long items. The box has five long panels, one of which fully overlaps. The ends also fully overlap.
- A Full Telescope Box has two fully telescoping sections. The sections may be formed by staples, die-cut locks, adhesive, etc.
- A Partial Telescope Box has two sections. The top telescopes partially over the bottom.
- A corrugated tray is often used for display purposes or used with a shrink wrap
- Corrugated corner pads can be used for product support and cushioning
- Special die-cut shapes have almost endless designs and uses.
- etc

Single-Face Laminate

A limitation of common corrugated material has been the difficulty in applying fine graphic print for informative and marketing purposes. The reasons for this stem from the fact that prefabricated corrugated sheets are relatively thick and spongy compared to the thin and incompressible nature of solid fibre paper such as paperboard. Due to these characteristics of corrugated; it has been mainly printed using a flexographic process, which is by nature a coarse application with loose registration properties.

A more recent development popular in usage is a hybrid product featuring the structural benefits of corrugated combined with the high-graphics print of lithography previously relegated to paperboard folding cartons. This application, generally referred to as 'Single-Face Laminate', begins its process as a traditional fluted medium adhered to a single linerboard (single-face), but in place of a second long-fibered liner, a pre-printed sheet of paperboard such as SBS (solid bleached sulfate) is laminated to the outer facing. The sheet can then be converted with the same processes used for other corrugated manufacturing into any desired form.

Specialized equipment is necessary for the material construction of SFL, so users may expect to pay a premium for these products. However, this cost is often offset by the savings over a separate paperboard sleeve and the labor necessary to assemble the completed package.

Recycling

Old corrugated containers are an excellent source of fiber for recycling. They can be compressed and baled for cost effective transport to anywhere in need of fiber for papermaking. The baled boxes are put in a hydropulper which is a large vat of warm water. Extraneous materials (string, films, strapping, tape, etc) are removed and contaminants are filtered and separated out. The pulp slurry is then used to make new paper and fiber products. Recycling helps resource conservation and helps developing countries without sustainable wood resources build a paper and packaging industry locally and develop their exports to global markets.

13.4 Carton

Carton is the name of certain types of containers typically made from paperboard which is also sometimes known as "cardboard".

Artistic Design

In art history, the carton (pronounced the French way) was a drawing on ordinary cardboard, used as life-size design for the manufacture in an atelier of a valuable Tapestry, such as a Gobelin. During the weaving it hung behind the tapestry in the making, a time-consuming process thus in a creative sense simplified to 'mechanical' painting-by-numbers.

As these were extremely valuable, often commanded by the very richest art-buyers, including princes who hung them in their palaces and even took them on their travels as prestigious displays of wealth, often with a visual message, especially the world-famous Flemish ateliers were deemed worthy to have cartons made by some of the greatest graphic artists of the time, including such celebrated painters as Rubens.

Packaging Types of Cartons

A carton is a type of packaging, generally for food. They come in many different shapes; milk cartons are upright boxes with spouts, egg cartons are long boxes with cups to hold the eggs upright.

Egg cartons or trays are designed to protect eggs while in transit. There are three types of raw materials which are used in the manufacture of egg cartons: recycled paper, Styrofoam and, more recently, PET.

In North America, the three largest producers of egg cartons are CKF Inc., Hartmann North America and Pactiv Corporation.

Packaging History

An early American packaging industry pioneer was the Kieckhefer Container Company, which was run by John W. Kieckhefer. The company excelled in the use of fiber shipping containers, which especially included the paper milk carton. In 1957, through an exchange of stock, the Kieckhefer Container Co. holdings were merged with the Weyerhauser Timber Company of Tacoma, Washington.

Shape

Although quite often shaped like a cuboid, it is not uncommon to find cartons lacking right angles and straight edges, as in surrounds used for ice cream. The number of corners on any given carton is a function of the product it contains. For example, a product with eight vertices would require a box also with eight corners. Exceptions to this rule are irregularly shaped products, like televisions, in which case polystyrene inserts are used to standardize the shape.

Materials

Despite the name, cartons can be made from many materials, and in fact are often a composite.

Many cartons are made out of a single piece of paperboard. Depending on the need, this paperboard can be waxed or coated with polyethylene to form a moisture barrier. This may serve to contain a liquid product or keep a powder dry.

Labelling

The labels on cartons generally contain information. Much of this information is considered useful by many people, because it is very informative. For example, many labels have featured the following:

- Label - indicative of the name, contents and producer of the carton.
- nutritional information
- contact instructions - how to contact the manufacturer in a complicated way involving hand-drawn facsimiles.
- Ingredients - exposes the public to terms like propylene glycol for the first time.
- contests - again, hand drawn facsimiles are involved. Self-addressed stamped envelopes are also often required.

13.5 Paperboard

Paperboard **is a paper**-like material, usually over ten mils (0.010 inch) in thickness. Some types of paperboard (corrugated medium and linerboard) are used in the construction of corrugated fiberboard. Other types of paperboard are used for folding cartons, set-up boxes, carded packaging, etc.

Cardboard is a generic, non-specific, lay term used to refer to any heavy paper-pulp based board, such as card stock, or to corrugated fiberboard.

Thickness, bulk and caliper are significant aspects to paperboard manufacture, as the end product must be able to withstand the bending forces of packaging, load carrying, shipping etc., while still maintaining their shape. Consequently, these paperboard products are usually comprised of two or more plies to provide the required stiffness.

Within the industry, paperboard is referred to by weight (pounds per 1000 square feet (MSF) or edge-crush pounds per lineal inch for stacking strength properties), pressure (pounds per square inch), by thickness (by thousandths of an inch, or 'points' - i.e. 0.009" thick paperboard is also known as 9-point) or by name (food board, corrugating medium, boxboard etc.).

Will Keith Kellogg first used paperboard folding cartons to hold flaked corn cereal, and later when he began marketing it to the general public, a heat-sealed waxed bag of Waxtite was wrapped around the outside of the box and printed with their brand name. This marked the origin of the cereal box, though in modern times the sealed bag is plastic and is kept inside the box rather than outside.

Today paperboard packaging in general, and especially products from certified sustainable sources, are receiving new attention, as manufacturers dealing with environmental, health, and regulatory issues look to renewable resources to meet increasing demand. It is now mandatory in many countries for paper-based packaging to be manufactured wholly or partially composed of recycled as well as tree-free fibers.

CHAPTER 14

Labelling of Pharmaceutical Packages

Summary

14.1 Introduction

"Labelling", as the term is used in the pharmaceutical industry, entails much more than the information that appears on a bottle or package. Labelling is a concise and comprehensive statement of the best information about a drug's quality, efficacy and safety.

14.2 Indian Rules Pertaining to Labelling

Drug & Cosmetic Act 1940 relating to Labelling of drugs and medicines.

14.3 Label Classification

Glued paper, Self Adhesive (Pressure Sensitive) Labels, Heat Seal Labels

14.4 Label Printing and Coding

Noncontact Coding

Two methods are now in use for marketing packages including drug containers in a way that does not depend on physical contact between the container and the marking machinery: ink jet printing and laser coding.

Ink Jet Printing: Laser Coding:

14.5 Bar Coding on Labels

Bar code is a computer coding system that uses a printed pattern of lines or bars to identify products, mail and packages, customer accounts, and the like. A bar code (often seen as a single word, *barcode*) is the small image of lines (bars) and spaces that is affixed to retail store items, identification cards, and postal mail to identify a particular product number, person, or location. A ***barcode reader*** is used to read the code.

14.6 Insert, Outsert and Bookserts

As noted earlier, the extensive information that must be supplied with some drugs requires the frequent use of inserts and outserts. As the names imply, inserts are printed matter that is inserted with in a box that also contains the primary drug package. The new concept in pharmaceutical Labelling in the introduction of "The pharma-booksert".

14.7 Mislabelling and Medication Errors

Mislabelling has been noticed as a major cause of medication error. During the production of a drug product, the one mistake that can occurs most consistently is applying the wrong label". Mislabelling is not only the most frequent problem; it is also the unsafe one because of the obvious potential for injury or death.

14.8 The Labelling Life Cycle

The overall life cycle of a label can be thought of as having three nested workflow cycles:

- The preparation cycle,
- The negotiation cycle and
- The post-release cycle.

14.9 Structured Product Labelling (SPL)

Since pharmaceutical companies must start submitting Labelling in Structured Product Labelling (SPL) to the FDA in 2005, this is not an academic question but a serious business decision. Structured information contains both content (words, pictures, etc.) and some indication of what role that content plays (for example, content in a section heading has a different meaning from content in a footnote, which means something different than content in a figure caption or content in a database table, etc.). Almost all documents have some structure.

14.1 Introduction

"Labelling", as the term is used in the pharmaceutical industry, entails much more than the information that appears on a bottle or package. Labelling is a concise and comprehensive statement of the best information about a drug's quality, efficacy and safety. Quality refers to the precise chemical composition of the drug, the strength and physical form in which it is supplied, and the rules for its storage and handling. Efficacy refers to the medical conditions for which it is indicated and the therapeutic effects of use, as well as the proper dosing for the accepted indications. Safety refers to potential side effects, contraindications and other consequences of use, including rules for monitoring patients with various special conditions.

Labelling is intended as a guide both for medical practitioners and for patients. It is presented as a textual narrative, but it also contains important detailed data elements.

In most countries the regulatory agencies have established rules for the content and format of Labelling, largely by means of a list of *pro forma* headings that defines the topics to be addressed, with additional rules for the exact information to be included for particular kinds of drugs. In the United States these are set out in the Code of Federal Regulations under Title 21 concerning The Food And Drug Act. 21 CFR 201.56 defines the major topical sections to be included as:

- Description
- Clinical Pharmacology
- Indications And Usage
- Contraindications
- Warnings
- Precautions
- Adverse Reactions
- Drug Abuse and Dependence
- Overdosage
- Dosage and Administration
- How Supplied

14.2 Indian Rules Pertaining to Labelling

The Indian rules relating to Labelling of drugs and medicines as per Drug & Cosmetic Act 1940 which includes the particular to be mentioned on the label in addition to the name and dose etc. of the drug are summarized as follows

I Drugs Needing Cautionary Labelling

Drugs	*Particulars*
Schedule G drugs	"Caution: It is dangerous to take this preparation except under medical supervision."
Schedule H drugs	(i) "Schedule H drug. Warning: To be sold on the prescription of a Registered Medical Practitioner only." (ii) Symbol R prominently on left hand top corner of the label. (iii) Symbol NR prominently on left hand top corner if drug is covered under Narcotic Drugs and Psychotropic substances Act.
Schedule X drugs	(i) "Schedule X drug. Warning: To be sold on prescription of RMPs only. (ii) Symbol XR in red on left hand top corner.
Schedule C drugs	(i) Where a test for maximum toxicity is prescribed a statement that it has passed that test. (ii) Nature and percentage of antiseptic, if any.
Veterinary drugs	(i) 'Not for human use. For animal treatment only, (ii) Head of any domestic animal.
Ophthalmic solutions/ Suspensions/ointments	(i) Use within one month of opening. Not for injection. (ii) Name and concentration of preservative. (iii) Words. (a) If irritation persists or increases discontinue use and consult physician. Keep container tightly closed.

	(b) Do not touch the dropper tip/other dispensing tip to any surface since this may contaminate. "Before fixing dropper there is a possibility of touching it. Scrub hands with soap and inse with water.
Ophthalmic ointments	Warning: If irritation persists or increases discontinue use and consult physician.

II Labelling Requirements for Other Drugs

Patent and proprietary	(i) Names and quantities of active ingredients.(ii) If vitamins are present, the following words as the case may be. "For therapeutic use" "For prophylactic use" "For pediatric use" and age of child/infant
Medicines for external use/ Medicines containing Methylated spirit	"FOR EXTERNAL USE ONLY" in capitals.
Mechanical contraceptives	Particulars specified in Schedule R.
Disinfectants	(i) Grade and phenol coefficient. (ii) Method of use.
Alcoholic preparations	Percentage of alcohol
Colored preparations	Name and percentage of color

III Labelling of Medicines for Export

The drugs meant for export are exempt from provisions of labelling except for their name, name and address of manufacturer, his license no. and batch No., date of expiry, if any, should also be indicated. The drugs may be labelled in accordance with the provisions of labelling of the country of export.

IV Labelling of Dispensed Medicines

Medicines dispensed on prescription of RMPs are also exempt from the normal provisions. However, following particulars are required to be included on their labels:

(a) Name and address of the supplier.

(b) Name of patient and quantity.

(c) Serial No. in prescription record.

(d) Dose, if it is for internal use and the words 'FOR EXTERNAL ONLY' if it is for external application.

14.3 Label Classification

Pharmaceutical labels can be classified into glued paper, self-adhesive or pressure sensitive, and heat sensitive categories.

1. **Glued Paper**

   It category is the oldest one which is rarely found in drug Labelling operations today because of conjunction created by glue with high speed automated equipment.

2. **Self Adhesive (Pressure Sensitive) Labels**

   A pressure sensitive label is a material, which has adhesive glue on the backing. When pressure is applied it will adhere to most surfaces.

   Self-Adhesive labels comprise of the printed label, paper, PVC or polyester. A layer of adhesive is permanently and insistently lackey when dry and a transparent silicone-coated backing sheet, which allows the adhesive coated label to be wound upon a roll. The adhesive layer, which is used to be solvent based is now usually a water based acrylic or rubber-based latex composition which can be adapted for degree of tack (which allows peelability if desired) water removing capability, low or high temperature adhesion, UV light resistance and general climate ability. Care must be exercised in choice of adhesive to avoid components that could travel through the wall of a plastic container and contaminate the drug package contents. At the same time, it is essential that the adhesive must be dominant enough to strongly resist label loss or switching that could cause an innocent consumer to take the incorrect drug.

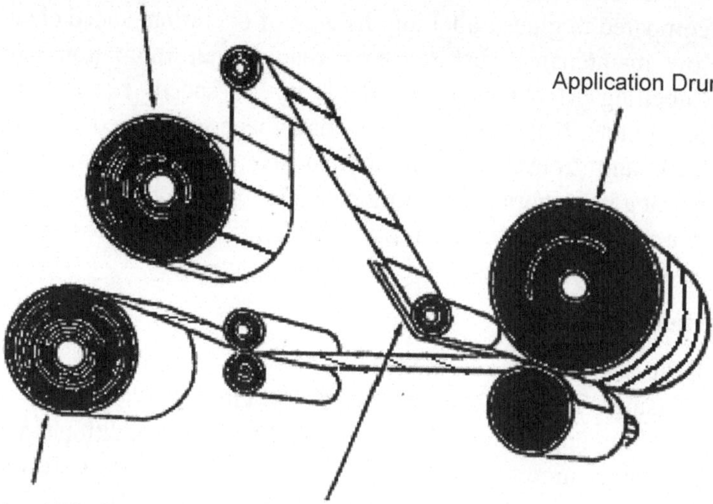

Fig. 14.1 Pressure sensitive Labelling machine.

Self-Adhesive or Pressure sensitive label supply is more or less made by manufacturers specializing in this activity. After these layers have been assembled, the paper and adhesive layer is dying cut to permit for afterward separation of the individual labels in the Labelling machine. These are then shipped in roll shape to the pharmaceutical industries.

The automatic machine that applies these label to the container consists of an relax station where the roll of label supply is fed around a stripper plate or peeler bar over which it reverses at direction of almost 180^0. This sharp bending action caused the backing sheet to peel away from the die, cut label array and freeing each separate label for pickup by the container (Fig. 14.1). The backing sheet can be separately wound up on another roll for final disposal. Each label is then either:

- Picked up by a vacuum head, positioned over the container, and pressed in place by an air jet which replaces the vacuum or
- Picked up by a container by its leading edge and moved past a pressure plate or plates which press the label tightly in place.
- Picked up by a vacuum drum and applied to the container in one continuous motion.

Although the costs of self adhesive label are roughly twice as compared to glues Labelling, its ease of operation, speed cleanliness and low maintenance features have made it far more popular than glue Labelling. Machines are available with a speeds exceeding 600 labels per minute. Dual head machines have been introduced which apply both back and front labels simultaneously. Pressure sensitive systems is gaining its importance in drug package Labelling over heat seal systems, due mainly to the comparative simplicity, effectiveness and operational ease of pressure sensitive Labelling equipment.

3. Heat Seal Labels

Heat seal labels are made by applying a combination of heat and pressure. These labels are used for tamper evident and permanent Labelling applications. Heat seal label have line additional advantage of providing more secure Labelling system. Since heat activated adhesive usually provide a stronger bond to the container wall.

Heat seal labels need the application of heat to activate the adhesive, which can be PVC, LDPE, or EVA copolymers. These adhesives when cool are not tacky and so require no backing sheet to enable the content labels to be wound on a roll. Thus the cost of these labels is usually lower equalize by the higher cost of the equipment required. For heat sensitive preparations, like some foods and medicines, the adhesive can be formulated in a so called delayed action manner in which the application of heat converts the adhesive to a pressure sensitive form so that heating equipment by no means comes near the manufactured goods being labeled. This mode has the further benefit that the heating elements need not have complex shapes to conform to irregular container shapes and surfaces.

The automatic machinery that applies heat seal label seal to containers consists of a station that holds the roll of labels from which they are cut off and fed to a heated vacuum drum. This drum contacts the container and rolls with it, pressing the label against the container wall. The pressure and dwell time required in this stop are both greater than in pressure sensitive Labelling. The alternative to this process is flagged on Labelling which is used for containers that are not cylindrical. Here the heated label is pressed on the container along a label press. The rest of the label is pressed on the container along a label edge. As the container those travels down the belt cold pads press the rest of the label in place. As with delayed action heat seal labels, this technique avoids heating the pressure pads, which must be specially contoured to fit the irregular container contour.

14.4 Label Printing and Coding

- **Noncontact Coding**

  Two methods are now in use for pharmaceutical packages including drug containers in a way that does not depend on physical contact between the container and the marking machinery: ink jet printing and laser coding. Both methods are being progressively used because they are independent of exterior texture and design. For example, an ink- jet printer can readily mark a ripe peach; a laser can mark a surface that will not accept inks. Both will place codes over rough and irregular surfaces.

- **Ink Jet Printing**

  In its simplest form, ink – jet printer ejects a continuous fine stream of ink through a nozzle and then through an ultrasonic device that breaks up the stream into a series of attached droplets. This stream is broken into individual droplets by mechanical and electrostatic forces as they pass through a positively charged tunnel. In the tunnel, the droplets are given varying degrees of electrostatic charge. The strength of which governs the extent of deflection they experience as they subsequently Passover a negatively charged defector plate. This varying deflection sets the pattern they form when they strike the surface to be printed. A computer program controls the charge applied to each drop so the printed pattern can be changed by a keystroke operation.

  This technique has the advantage of great mechanical simplicity and is used in drug packaging for lot numbering expiration date coding and consecutive numbering. It is particularly well suited for such used because the information printed can be programmed to be different for each container that the printer senses without incurring the cost and complexity that would be involved were label to be uniquely printed by a converter and then later mated to the correct container. Obviously, this technique is applicable to applying bar codes on containers and cartons.

- **Laser Coding**

  A laser is a beam of tightly collimated light waves created by a high voltage charge of electricity passed through a tube containing various gases. When this energetic beam strikes a paper, plastic or painted surface, the paper plastic or paint is instantly vaporized. If the subsurface is laser resistant material such as metal, it remains, making the contrast between the etched and un-etched surface clearly evident. A pattern can then be created bypassing the beam through a bars stencil and then focusing it before it strike the surface to be printed.

Since this technique is very fast. It is readily adapted to creating sharp, no blurred images at container line speeds in excess of 600 per minutes. The imprinted code is more permanents than ink and can consist of characters as small as one – sixteenth of an inch. Since a laser beam is light mirrors can be used to alter its direction and it can be focused with lines the system are highly reliable can require almost no operator attention. To date they are little used for pharmaceutical package Labelling. Cost being the major deterrent.

14.5 Bar Coding on Labels

Bar code is a computer coding system that uses a printed pattern of lines or bars to identify products, mail and packages, customer accounts, and the like. A bar code (often seen as a single word, *barcode*) is the small image of lines (bars) and spaces that is affixed to retail store items, identification cards, and postal mail to identify a particular product number, person, or location. The code uses a sequence of vertical bars and spaces to represent numbers and other symbols. A bar code symbol typically consists of five parts: a quiet zone, a start character, data characters (including an optional check character), a stop character, and another quiet zone.

A *barcode reader* is used to read the code. The reader uses a laser beam that is sensitive to the reflections from the line and space thickness and variation. The reader translates the reflected light into digital data that is transferred to a computer for immediate action or storage. When the code is read by an optical scanner linked to a computer, the computer can provide and record information about the item, such as its price or the quantity sold, from and to databases.

Bar codes and readers are most often seen in supermarkets and retail stores, but a large number of different uses have been found for them. They are also used to take inventory in retail stores; to check out books from a library; to track manufacturing and shipping movement; to sign in on a job; to identify hospital patients; and to tabulate the results of direct mail marketing returns.

Most drug packagers now bar code some of their products since bar coding offer many advantages:

- Input and collection of transaction data is automated. Eliminating the error potential of manual keystroke entries.
- Efficient inventory management is enhanced.
- Manufactures processes that depend on positives identification of container contents become faster and more accurate and Labelling errors can be reduced.
- Both wholesalers and pharmacies gain efficiency in the entire range of individual operations that take place when a drug goes in and out of these institutions.

The National Drug Code (NDC) has become a useful device for identifying each drug product with a unique, all numeric number as opposed to numbers and letters that identities the drug manufactures / labeler, product and package size. Drug manufactures offer and drug wholesaler order, products using the NDC, pharmacy computer systems drug industry support systems, and third party prescription claims processing, sales tracing and reporting all use the NDC as a basis of identifying, describing and paying for pharmaceutical services. The law does not yet require the use of the NDC, nut virtually every drug product label contains such a number.

USP is now considering recommending the use of a three-character identification code for solid dosage forms of drugs, vitamins and minerals. This code would be printed on individual tablets or capsules and would simply identify the drug, not the manufacturer or the strength, USP argues an important consumer benefit when a prescription is refilled in using a generic equivalent instead of a proprietary product the buyer could verify the equivalence of the generic by comparing the codes, which must be the same for both.

Both the NDC and its bar equivalent are too large to be printed on individual tablets or capsules. Nevertheless, USP proposed to establish a linkage between the two systems that can be tracked by a computer if necessary. Responses from manufacturers have expressed concerns of one kind or another, while responses from consumers and consumer groups were generally favorable. The proposal is being altered to reflect industry concerns and will likely be adopted soon on a voluntary basis. The widespread use of the NDC provides a logical basis for the adoption of bar coding to identify drug products using the Universal Product Code format. A bar code is a series of black and white bars and space that represent a series of characters or symbols. Its purpose is to code information in a form that is readily read by a machine that sweeps a small spot of light across the symbol, which must be large enough so that the light cannot wander outside its limits. Various formats have been adopted but the Universal Product Code (UPC) is appropriate to capture the information contained in the NDC, which covers most of the essential information needed to verify that a product has been correctly labeled.

The only significant negative is the space that a functional bar code requires on an already crowed label. This problem becomes particularly acute in the case of unit dose labels, so mush so that only recently has any supplier bar coded unit dose packages on each dose. Although a stacked bar code would probably be best for very small packages like blisters, adoption of that system is still in the future since scanner and verifier changes are necessary. Thus the first bar coded blister used code 128; the smallest possible symbol is compatible with existing readers. It is so small that only the NDC number can be included – the expiration date and lot numbers are missing.

Bar coded unit dose packages have many advantages in hospitals, where the nurse could scan a bar code on the patient's wrist band and the bar code on the package to verify that the medicine was the correct one, some hospitals have reported impotent savings when they applied bar codes to blisters. Bar coding by the manufacturer would certainly be preferable to the hospitals doing it themselves.

In *July, 2001* The National Coordinating Council for Medication Error Reporting and Prevention (Council) (NCC MERP) has issued a set of recommendations that calls upon the U.S. Food and Drug Administration (FDA) and the U.S. Pharmacopoeia (USP) to collaborate with pharmaceutical manufacturers and other appropriate stakeholders to establish and implement uniform bar code standards, including standards for unit-of-use packages.

The Recommendations state that patient safety can be improved by information technology through the use of machine-readable codes such as bar codes in a standardized format. A scannable bar code can help guarantee that the right drug and the right dose are being administered to the right patient. Technology developments allow for increased information to be imbedded within a bar code and make the coding of smaller packages possible.

The Council recommends that the data elements of the bar code include, at a minimum, the National Drug Code (NDC), the lot/control/batch numbers, and the expiration date. The Council also recommends that the data elements be uniformly ordered, and that the bar code print density should be consistent to enable an accurate scan each time.

14.6 Insert, Outsert and Bookserts

Labelling documents often run to more than ten pages. They are usually distributed as a leaflet accompanying each package, and are therefore often referred to as a *Package Insert* and *Outserts*. As the names imply, inserts are printed matter that is inserted with in a box that also contains the primary drug package. Outserts are strange but highly descriptive words that must have been coined by a pharmaceutical drug packager, consist of the same printed matter that is attached to the primary container types discussed above by double sided tape or by plastic banding or over wrap, usually with a shrink film.

The new concept in pharmaceutical Labelling in the introduction of "The pharma-booksert". The pharma-booksert relies on easy to read product information for the patient and provides the mechanisms for the pharmaceutical company to identify their end-user, thus allowing for on-going support to the patient as needed. The booksert is printed in multiple colors with inclusive patient information and professional detailing required for sample packages and starter kits. It even offers a direct mail piece providing privacy to the patient searching for more information on their medication.

Before the days of plastics, boxes some times with dividers that helped avoid breakage of glass containers were common. The advent of unbreakable plastic containers and the concept of outserts combined to dramatically shrink the frequency of box use and lower packaging cost significantly.

After the editorial matter has been approved by the FDA, the insert / outserts is printed and delivered, either precut and folded or on a roll to the packaging line which feeds the insert to an online cutting and folding step. Since the law regards the information on insert/outsert to be as essential as the information on labels, the possibility of error while mating the insert/outsert with container must be reduced as far as possible and with the same dedication that is applied to elimination of Labelling errors. For this reason, roll- fed rather than precut inserts and outsert, are preferred.

An elegant way to handle these complexities is the so- called fix- a form technique recently from the U.K. hare the outsert is printed is small, crisp type on thin paper, fan – folded and adhered to the center of a flat, pressure sensitive label by extending the outer panel and using this extension to hold the adhesive. The combined label is spooled and fed to the Labelling machine and attached to the container just as if it were a simple pressure sensitive label. If both the parts of the label are bar coded, scanners can ensure the right outsert is mated with right label. Spooling minimizes the chance of mislabelling in the pharmaceutical plant. The pharmacist generally peels off the outsert, exposing an area that can carry the information he needs to provide.

14.7 Mislabelling and Medication Errors

"A medication error is any preventable event that may cause or lead to inappropriate medication use or patient harm while the medication is in the control of the health care professional, patient, or consumer. Such events may be related to professional practice, health care products, procedures, and systems, including prescribing; order communication; product Labelling, packaging, and nomenclature; compounding; dispensing; distribution; administration; education; monitoring; and use."

Mislabelling has been noticed as a major cause of medication error. During the production of a drug product, the one mistake that can occurs most consistently is applying the wrong label". Mislabelling is not only the most frequent problem; it is also the unsafe one because of the obvious potential for injury or death.

Largest source of pharmaceutical product recalls is due to copy related errors on cartons, labels or inserts. In fact, over 51% of all product recalls in a six-month period during 2005 were the result of mislabelling and seal failures according to statistics.

Some of the mislabelling situations that prompted class I recalls (the most serous category) may be understood by followings:

- Containers of camphorated sprits intended only for external use were labeled as castor oil.
- Containers that are labeled quinidine sulfate and intended for use in control of irregular heartbeat actually contained dicyclomine hydrochloride, an antispasmodic used to relive intestinal cramps but totally ineffective in controlling life – threatening irregular heartbeat problems.

- A diuretic preparation was mislabeled as a calcium channel-blocking agent used to relieve angina.
- Bottles contained an antibacterial drug used to treat intestinal parasites, mislabeled as propanolol hydrochloride, a heart acting drug.

14.8 The Labelling Life Cycle

Drugs evolve over their market life: new product variants are developed, new uses are found, and new safety information emerges. Of all submission documents, Labelling is the most frequently updated. Labelling updates occur throughout the market life of a drug; many average two to three substantive updates per year.

The overall life cycle of a label can be thought of as having three nested workflow cycles as represented in Fig. 14.2.

- **The preparation cycle**: the cycle of authoring, review and approval within the sponsor company;
- **The negotiation cycle**: the cycle of review and approval conducted between the FDA and sponsor; and
- **The post-release cycle**: during which new quality, efficacy or safety information is acquired that triggers an update to the label.

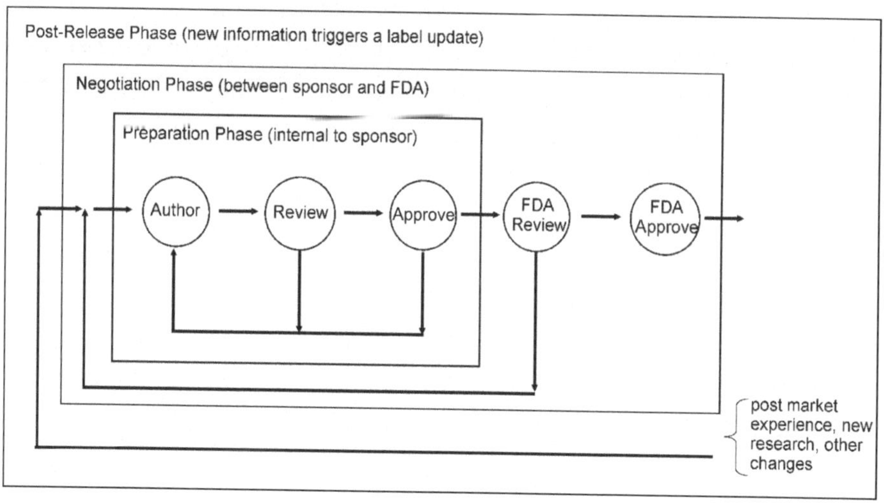

Fig. 14.2 Label life cycle.

14.9 Structured Product Labelling (SPL)

In most cases, information technology is applied primarily to increase internal efficiency to the benefit of the host organization or industry, though it is often touted as being ultimately to the benefit of the consumer.

A long-standing problem, exacerbated by an aging population and a continually expanding repertoire of medical interventions, is the emergence of new efficacy and safety information about drugs already on the market. A drug, used properly to treat an indication for which it has been tested, may show unexpected side effects or significantly altered therapeutic value when used in the context of other conditions or treatments not considered, or even known, at the time of its development. Sponsors and regulators operate systems to collect post-marketing adverse effect information in an effort to identify the signal of a real effect amid the noise of mere coincidence. Once detected, the effect and its remedy need to be communicated to the medical community and patients by means of updated Labelling.

Current methods of distributing new Labelling use a combination of print and electronic media. The electronic channels include the sponsor's web site, the FDA web site, and services operated by publishers and by medical and patient-oriented organizations. Print information can take many months to work its way through the channels to the practitioner and patient; electronic information is available more quickly, but the sites in question may not be updated synchronously and vary widely in their comprehensiveness and search facilities. For example, a sponsor's site can be expected to carry up-to-date Labelling for its products, but only its products, while *Drugs@FDA*, the FDA's repository of all approved Labelling, is searchable only by product name, which is sufficient for the practitioner with a specific product in mind, but less useful when the goal of the search is to identify alternative products in the context of given indications and contraindications. In addition, the formats used do not include a standard method for indicating version-to-version changes.

Since pharmaceutical companies must start submitting Labelling in Structured Product Labelling (SPL) to the FDA in 2005, this is not an academic question but a serious business decision. Structured information contains both content (words, pictures, etc.) and some indication of what role that content plays (for example, content in a section heading has a different meaning from content in a footnote, which means something different than content in a figure caption or content in a database table, etc.). Almost all documents have some structure.

The primary purpose of SPL is to support electronic distribution and electronic update over the life cycle of the drug. SPL is to be the basis of a broader initiative that will include automated distribution of Labelling content from the sponsor, through the FDA, to a central repository operated by the National Library of Medicine as shown in Fig. 14.3.

The availability of an authoritative, up-to-date central repository for Labelling, organized and searchable on a variety of terms, will benefit consumers largely by reducing the possibility of adverse effects. But this is only the first step in the use of SPL. There are initiatives underway within the medical community, with the support of the U.S. Department of Health and Human Services, to promote and standardize e-prescribing. It is likely that the coded data elements will be used in support of electronic prescribing.

While the pharmaceutical companies must bear an initial cost and on-going costs to create, maintain and distribute Labelling in SPL for the benefit of the consumer, timely information distribution may also serve to reduce their potential product liability.

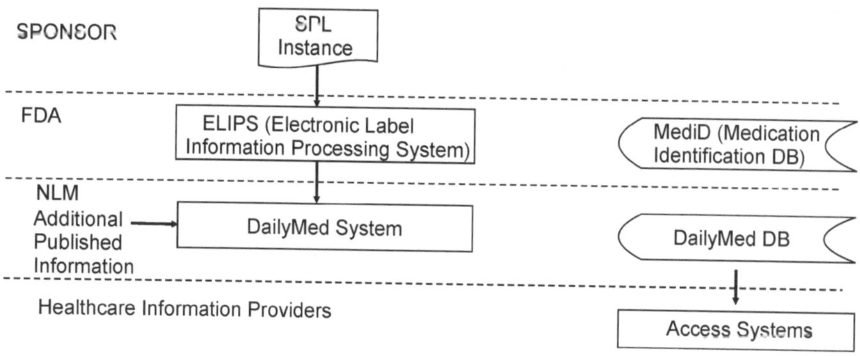

Fig. 14.3 Planned SPL distribution.

If SPL standards are applied only for document interchange, then the content can be encoded in XML (XML is a markup language for documents containing structured information) as a post-production step without impacting established document management practices or the existing authoring, review and approval process. While apparently simple, this approach has some drawbacks and does not exploit the potential of XML to improve the process itself. On the other hand, any changes to a complex and critical business process must offer benefits that outweigh the risks and costs.

CHAPTER 15

Regulatory Aspect of Pharmaceutical Packaging

Summary

15.1 Introduction

The life saving industries especially pharmaceutical industry is one of the most regulated industries. The industry requires making control on various aspects like:

- Good Manufacturing practice (GMP)
- Good laboratory practice (GLP)
- Good clinical practice (GCP)
- Labelling instructions
- Control on how the product is advertised (mainly prevalent in the regulated market), and
- Product disposal regulations.

The European Union (EU) has been fairly successful in the pharmaceutical area while achieving:

- Centralized system in place,
- Decentralized system in place,
- Harmonization of format,
- Increase acceptance outside EU, and
- Harmonization of standards

The various European Union Guidelines Applying to Packaging are

- CPMP List of allowed Terms (III/359/91)
- Notice to Applicants (Updated 1998)
- Plastic Container Guidance (III/9090)
- Plastic in Contact with Food Directive (90/128)

15.2 FDA Packaging Guidelines

This guidelines defines the types of containers to be used, dividing them into parenteral (glass/ plastic) or non- parenteral containers (glass, plastic and metal) pressurized containers and bulk containers for active ingredients and drug products. Also listed are closures including, child – resistant and tamper – evident closure, liner, seal and elastomers when used for closure.

15.3 Legal Requirements for Labelling and Packaging

When the FDA evaluates a drug, the agency must be firmly convinced that the package for a specific drug will preserve the drugs efficacy as well as stability, identity, purity, strength and quality for its entire shelf life. Under the provisions of, the federal food, drug and cosmetic act 1940, it is the responsibility of the manufacturer to provide the safety of a packaging material and to get approval before using it for any food and drug products.

15.1 Introduction

FDA ensures the quality of drug products by carefully monitoring of drug manufacturers adhere to pharmaceutical regulatory compliance as per the current Good Manufacturing practice (cGMP). The cGMP regulations for drugs contain minimum requirements for the methods, facilities, and controls used in manufacturing, processing, and packaging of a drug product. The regulations make sure that a product is safe for use, and that it has the ingredients and strength it claims to have. Pharmaceuticals are perhaps the most strictly regulated products, subjected both to broad pre-marketing testing and constant post-marketing evaluation.

Pharmaceutical industry is one of the most regulated industries, controlled by rules and regulations and ethics of governing bodies. It requires making control on various aspects like:

- Good Manufacturing practice (GMP)
- Good laboratory practice (GLP)
- Good clinical practice (GCP)
- Labelling instructions
- Control on way, the product is advertised
- Product disposal regulations.

Gaining an understanding of the principles of pharmaceutical quality assurance and control, which have become increasingly important due to pharmaceutical regulatory interest. The following characteristics are the common requirement of most regulatory agencies:

1. Product/ Preparation related requirements:
    - Protection of the product
    - Protection of the consumer
    - Control of doses
2. Label related requirements
    - Information to the receiver
    - Legal control of the product
3. Environmental aspects
    - Packaging wastage
    - Ozone depletion
4. Consumer protection
    - Child- resistant closures
    - Tamper-evident packaging

Pharmaceutical packaging is a skill-based industry directly related with marketing of product. This industry is one of the fastest growing industries of India. The material like glass used in pharmaceutical packaging was an inert vessel with slight potential interaction with the product. With the introduction of plastics, there is much more scope for interactions between the product and the package. With more and more tailor made packages, the use of plastic is on an ever increase. They, however, pose problems such as extraction of material from the plastic. This is where the regulatory authorities become much more interested in the selection, composition and performance of the package. To with – stand the ever-increasing cost of research and development, the need for becoming a global player in pharmaceutical marketing is increasing. The regulatory for over 150 countries may have differing requirements. The solution for this lies in the realization of the fact that out of the total market share. European Union (EU), represents 30% of the market, the US approximately 27% and Japan nearly, 18%. These add up to 75% of the market share. International Conference on Harmonization (ICH) between these areas on certain issues may lead to a further improvement in the situation.

The EU has been fairly successful in the pharmaceutical area while achieving:

- Centralized system in place,
- Decentralized system in place,
- Harmonization of format,
- Increase acceptance outside EU, and
- Harmonization of standards

The various EU guidelines applying to packaging are

- CPMP List of allowed Terms (III/359/91)
- Notice to Applicants (Updated 1998)
- Plastic Container Guidance (III/9090)
- Plastic in Contact with Food Directive (90/128)

15.2 FDA Packaging Guidelines

FDA packaging guidelines define the types of containers to be used, dividing them into parenteral (glass/ plastic) or non- parenteral containers (glass, plastic and metal) pressurized containers and bulk containers for active ingredients and drug products. Also listed are closures including, child-resistant and tamper-evident closure, liner, seal and elastomers when used for closure. The packaging components are discussed for physical, chemical and biological characteristics, specifications and tests to be applied, stability and compatibility. The various tests mentioned have been discussed separately under container testing.

According to FDA guidelines, for submitting document for packaging for human drugs and biological the followings are required.

Purpose

- Package must maintain standards, identity, strength, quality and purity of drug for intended shelf – life.
- Full information needed.
- Type of container / closure.
- Suitability for intended use.
- Submission of packaging information and date

Environmental Concerns

With increased environmental concern there has been a considerable pressure to reduce contamination of the environment, with particular concern on amount of packaging and its disposal. Ozone depletion is also of concern with the use of pressurized containers. Regarding this aspect, the increase in concern has led to the

Along with original form of any drug following information are to be supplied	
Descripton	General Description of container with closure Name, product code, physical description, manufacturer, raw material, any additional treatment
Suitability	Protection against light, oxygen, moisture leakage, microbes, dirt, etc. Safety – chemical composition extractable material test as per pharmacopoeia or any other relevant test *Compatibility – test* reports for packaging interacting with product of vice versa. Performance – suitability for its intended use e.g. MDI Quality control. Various tests and acceptance limits. Dimensional check.
Stability	Expiry of Pharmaceutical Product / Formulation
Performance	Consistency in composition Manufacturers protocol for release and manufacturers process of component

European E Commission packaging waste directive, which requires:

- Reduction in quantity of waste,
- Reduction in harmfulness of waste,
- Increase in reuse of packaging,
- Recycling and recovery of packaging waste, and
- Reduction of the total packaging to be disposed of.

About one third of cost of medicine is accounted for the packaging materials. For many years the role of packaging was undermined, but now the importance of pharmaceutical packaging has been duty recognized. The packaging not only provides the protection to the product from contamination external influences and physical damage but also helps in identification of the product and display correct information. The packaging material and product must be compatible. More care should be exercised in pharmaceutical packaging because any failure could result in contamination or loss of potency of the drug leading to failure of therapy, to adverse effects, to injury or even death of the patient, Maintenance of shelf – life of product is the foremost requirement.

As whole plastics, glass, metal and paper are the most commonly used packaging material in pharmaceutical technology. Each one has its own pros and cons. The selection of the packaging material of the drug is mainly dependent on the nature of the product, level of protection required, stability, cost, machinability, shipment requirement and dispensing of the dosage from. Basically the pharmaceutical packaging is classified as primary packaging (bottles, jars, cans, strips, blister etc), Secondary packaging (cartons, corrugated boxes etc.) and accessories (caps, closures and labels). Packaging further needs to be tamper-resistant, child-resistant and customer friendly. Stringent quality control test should be practiced to check the quality of not only the raw packaging material but also the finished products. It is forecasted that US pharmaceutical packaging market is to reach $ 26.4 billion by 2009. Such growth will lead to the development of value added containers and accessories. Development of patient friendly packaging, anticipation and preparation of future legislative direction will be the need of coming years.

Importance of drug delivery system has been recognized and it is capturing a good market share. This will grow further in future. Packaging requirement of novel drug delivery systems are different form those of the conventional dosage forms. The innovations and developments in packaging of such product should be foresighted and will be a key factor in successful marketing of these products.

15.3 Legal Requirements for Labelling and Packaging

When the FDA evaluates a drug, the agency must be firmly convinced that the package for a specific drug will preserve the drugs efficacy as well as stability, identity, purity, strength and quality for its entire shelf life. Under the provisions of, the federal food, drug and cosmetic act 1940, it is the responsibility of the manufacturer to provide the safety of a packaging material and to get approval before using it for any food and drug products.

The FDA does not approve containers as such, but only the materials used in the container. A list of substances considered 'generally recognized as safe' (GRAS) has been published by the FDA. In the opinion of the qualified experts they are safe under specified conditions, assuming they are of good commercial quality

A material that is not included under GRAS or prior sanctioned, and is indented to be used with food, must be tested by the manufacturer, and the data must be submitted to FDA.

The rules and regulations for labelling and packaging under the Drug and cosmetic act 1940 are given in the part IX of the act. According to act labelling and packing of drugs other than Homoeopathic medicines is as under:

Rule 94. Exemption of certain drugs from certain provisions of this Part. – (1) Labels on packages or containers of drugs for export shall be adapted to meet the specific requirements of the law of the country to which the drug is to be exported but the following particulars shall appear in a conspicuous position on the innermost container in which the drug is packed and every other covering in which that container is packed:

(a) name of the drug;

(b) the name, address of the manufacturer and the number of the license under which the drug has been manufactured;

(c) batch or lot number;

(d) date of expiry, if any;

[Provided that where a drug, not classified under Schedule F, Schedule F(1) and Schedule X, blood products, Narcotic and Psychotropic Substances is required by the consignee to be not labeled with the name and address of the manufacturer, the labels on package or containers shall bear a code number as approved by the licensing authority mentioned in rule 21.]

[The provisions or rules 96 to 101 inclusive, shall not apply to a medicine made up ready for treatment, whether after or without dilution, which is supplied on the prescription of a registered medical practitioner provided that –

(i) the medicine is labeled with the following particulars: -

(a) the name and address of the supplier;

(b) the name of the patient and the quantity of the medicine;

(c) the number representing serial number of the entry in the prescription register;

(d) the dose, if the medicine is for internal use;

[(e) The words 'For External use only' shall be printed on the label if the Medicine is for external application;]

(ii) Condition (3) of the conditions in rule 65 is satisfied.]

Rule 95. Prohibition of sale or distribution unless labeled: Subject to the other provisions of these rules, no person shall sell or distribute any drug (including a patent or proprietary medicine) unless it is labeled in accordance with these rules.

Rule 96. Manner of Labelling: (1) Subject to the other provisions of these rules, the following particulars shall be either printed or written in indelible ink and shall appear in a conspicuous manner on the label of the innermost container of any drug and on every other covering in which the container is packed, namely:-

(i) The name of the drug:

[(A) For this purpose], [the proper name of the drug shall be printed or written in a more conspicuous manner than the trade name, if any which shall be shown immediately after or under the proper name and shall be]

(a) for drugs included in Schedule F or Schedule F(1), the given therein.

(b) for drugs included in the Indian Pharmacopoeia or the official Pharmacopoeias and official compendia of drug standards prescribed in rule 124, the name or synonym specified in the respective official pharmacopoeias and official compendia of drug standards followed by the letter 'L.P.' or, as the case may be, by the recognized abbreviations of the respective official pharmacopoeia and official compendia of drug standards;

(c) For drugs included in the National Formulary of India, the name or synonym specific therein followed by the letters 'N.F.I.';

(d) For other drugs, the international non – proprietary name, if any, published by the World Health Organization or where an international non – proprietary name is not published the name descriptive of the nature of origin of the substance.

(i) Omitted.

(ii) A correct statement of the net contents in terms of weight, measure, volume, number of units of contents, number of units of activity, as the case may be, and the weight, measure and volume shall be expressed in Metric system.

(iii) The content of active ingredients –

The shall be expressed –

(a) for oral liquid preparations in terms of the content per single dose, the dose being indicated in 5 milliliters

[Provided that where the dose is below 5 milliliters the contents of active ingredients may be expressed in terms of one milliliter, [3][or fraction thereof];

[Provided further that where the single dose is more than 5 milliliter, the content of active ingredients shall be expressed in terms of minimum single dose as approved by the licensing authority,]

(b) for liquid parenteral preparations ready for administration, in terms of 1 milliliter of percentage by volume or per dose in the case of a single dose container:

[Provided that if the preparation is contained in an ampoule it will be enough if the composition is shown on the label or wrapper affixed to any package in which such ampoule is issued for sale;]

(c) for drugs in solid form intended for parenteral administration in terms of units or weight per milligrams or grams;

(d) for tablets, capsules, pills and the like, in terms of the content in each tablet, capsule, pill or other unit, as the case may be;

(e) for other preparations, in terms of percentage by weight or volume or in terms of unitize per gram or milliliter as the case may be;

[Provided that clause (ii) shall not apply to a pharmacopoeial preparation where the composition of such preparation is specified in the respective pharmacopoeia and to a preparation included in the National Formulary of India].

(iv) The name of the manufacturer and the address of the premises of the manufacturer where the drug has been manufactured:

[Provided that if the drug is contained in an ampoule or a similar small container, it shall be enough if only the name of the manufacturer and his principal place of [manufacture] is shown.]

(v) A distinctive batch number, that is to say, the number by reference to which details of manufacture of the particular batch from which the substance in the container is taken are recorded and are available for inspection, the figure representing the batch number being preceded by the words 'Batch No.' or 'B No.' or 'Batch' or 'Lot No.' or 'Lot'

Notes. – 1. In the case of drugs manufactured by a continuous process, like Manufacture.

of magnesium sulphate, pharmaceutical chemicals, etc., the production resulting in one homogeneous mix of the finished products shall be considered as one "Batch":

2. In the case of power, liquid orals, ointments, etc., one "Batch Number" shall be assigned to all the containers field from one homogeneous bulk.

3. In the case of tablets, capsules, lozenges, torches, etc. one "Batch Number" shall be assigned to the products manufactured from one homogeneous mix ready for compression or filling.

4. In case of parental preparations sterilized by steam under pressure, one "Batch Number" shall be assigned to all containers filed from one homogeneous bulk solution and sterilized load.

5. In the case of containers of parental preparations filed from one homogeneous bulk solution and sterilized in more than one sterilizer load, the "Batch Number" assigned to the containers in the different sterilizer loads shall be same Batch Number" as is assigned to the homogeneous bulk solution,

[provided the samples taken from all the sterilizer loads pass the sterility test, and kept separate from one another unit the report of the sterility test is available.]

Explanation: For the purpose of chemical and other tests, representative samples from all containers filled from the homogeneous bulk solution should be taken.

6. In the case of parentral and other sterile products filled aseptically a "Batch Number shall be assigned to all containers filled from one homogeneous mix during one filling operation, the filling operation being completed in a period of not more than a day during which no schedule change in the filling assembly is made.

When containers are filled from one homogeneous mix, in a number of filling operations, the "Batch Number" assigned to the containers filled in individual filling operations shall be the same "Batch Number" as is assigned to the homogeneous mix, provided the samples taken from all the direction filling operations pass the sterility tests, and are kept separate from one another until the report of the sterility tests is available.

Explanation: For the purpose of chemical and other tests, Representative samples from all containers filled from the homogeneous mix should be taken.

7. In the case of medicinal gases produced by a continuous process of operation a week's production from one tank load shall be considered as a Batch.

(vi) Every drug manufactured in India shall bear on its label the number of the licence under which the drug is manufactured, the figure representing the manufacturing licence number being preceded by the words 'Manufacturing Licence Number' or 'Mfg. Lic. No.' 'M.L.'

(vii) Drugs specified in schedule P and their preparations including combinations with other drugs shall bear on their labels the date of manufacture and the date of expiry of potency, and the period between the date of manufacture and the date of expiry shall not exceed that laid down in the said schedule [1][under the conditions of storages specified there in.] [2][Drugs and their] preparations not included in schedule P, shall bear on their labels the date of their manufacture and also the date of their expiry which shall not exceed sixty months from the date of manufacture]:

[provided that this period may be extended by the licensing authority specified in clause (b) of rule 21 in respect of any specified drug if satisfactory evidence is produced by the manufacturer to justify such an extension.]

(viii) Drugs specified in schedule C(1) and their preparations including combinations with other drugs shall bear on the labels (a) the date of manufacture, (b) date of expiry of potency fixed by the manufacturer, and (c) where such drugs are imported, also the number of license under which the drug is imported, preceded by the words 'Import License':

[Provided that drugs in bulk form included in schedule C (1) which are not ready for use and not included in schedule P need not bear on the label the date of expiry of potency:]

Provided further that no reference shall be made to any other license number granted by any authority outside India on any label or container or in any covering in which the container is packed or in any other matter or advertisement enclosed therewith.

(ix) Every drug intended for distribution to the medical profession as a free sample shall, while complying with the labelling provisions under clauses (i) to (viii), further bear on the label of the container the words 'Physician' s sample – Not to be sold' which shall be overprinted.

(x) If any preparation contains not less than 3 per cent by volume of alcohol the quantity of alcohol shall be stated in terms of the average percentage by volume of absolute alcohol in the finished products.

(xi) In addition to the other particulars which are required to be printed or written under these rules, the label of innermost container of the following categories of drugs and every other covering in which the container is packed shall bear a conspicuous red vertical line on the left side running throughout the body of the label which should not be less than 1mm in width and without disturbing the other conditions printed on the label under these rules, namely: -

Narcotic analgesics, hypnotics, sedatives, tranquillizers, corticosteroids, hormones, hypoglycemic, antimicrobials, antiepileptic, antidepressants, anticoagulants, anti-cancer drugs and all other drugs falling under schedules G, H and X whether covered or not in the above list:

[Provided that the provisions of this clause shall not apply to –

(a) Preparations intended for animal treatment;

(b) Preparations intended for external use;

(c) Ophthalmic preparations and ear drops; and

(d) Sterile preparations such as sutures, surgical dressings and preparations intended for parenteral use.]

2. (i) the particular to be printed or written on the label of a mechanical contraceptive shall be as specified in schedule R.

(ii) The following particulars, in addition to those specified under sub – rule (1) shall be either printed or written in indelible ink and shall appear in a conspicuous manner on the label of the innermost container and on every other covering in which the container of a contraceptive, other than a mechanical contraceptive, is packed, namely:

(a) the date of manufacture;

(b) the date up to which the contraceptive is excepted to retain its properties;

(c) the storage conditions necessary for preserving the properties of the contraceptive up to the date indicated in sub- clause (b):

[Provide that for oral contraceptives it shall be sufficient to display on the label of the container the date of manufacture only.]

3. (i) The particulars prescribed in sub-rule (1) shall be printed or written in indelible ink either on the label borne by a container or vaccine lymph or on a label or wrapper affixed to any package in which the container is issued for sale. The said particulars shall be indelibly marked on the sealed container of surgical ligature or suture or printed or written indelible ink on a label enclosed therein.

(ii) Nothing in these rules shall be deemed to require the labelling of any transparent cover or of any wrapper, case or other covering used solely for the purpose of packing, transport of delivery.

4. (i) Where by any provision of these rules any particulars are required to be displayed on a label on the container such particulars may, instead of being displayed on a label, be etched, painted or otherwise indelibly marked on the container.

[Provided that, except where otherwise provided in these rules, the name of the drug or any distinctive letters intended to refer to the drug shall not be etched, painted or otherwise indelibly marked on any glass container other than ampoules.]

Explanation: For the purpose of this rule, the date of expiry shall be in terms of month and year and it shall mean that the drug is recommended till the last day of the month. The date or expiry shall be preceded by the words 'Expiry date'

Rule 97. Labelling of Medicines

1. The container of a medicine for internal use shall-
   (a) If it contains a substance specified in schedule, G is labeled with the words 'Caution: it is dangerous to take this preparation except under medical supervision' – conspicuously printed and surrounded by a line within which there shall be no other words;
   (b) If it contains a substance specified in schedule H be labeled with the symbol Rx and conspicuously displayed on the left top corner of the label and be also labeled with the following word: -

      'Schedule H drug – warning: To be sold by retail on the prescription of a registered medical practitioner only';
   (c) If it containing a substance specified in schedule H and comes within the purview of the [1][Narcotic Drugs and Psychotropic substances Act, 1985 (61 of 1985)] be labeled with the symbol NRx which shall be in red and conspicuously displayed on the left top corner of the label, and be also labeled with the following words: -

      'Schedule H drug – Warning: To be sold by retail on the prescription of a Registered Medical Practitioner only';
   (d) If it contains a substance specified in schedule X, be labeled with the symbol XRx which shall be in red conspicuously displayed on the left top corner of the label, and be also labeled with the following words: -

      'Schedule X drug – Warning: To be sold by retail on the prescription of a Registered Medical Practitioner only';
2. The container of an embrocating, liniment, lotion, [ointment, antiseptic cream,] liquid antiseptic or other liquid medicine for external application shall be labeled with the words in capital 'For External use only'.]
3. The container of a medicine made up ready only for treatment of an animal shall be labeled conspicuously with the words 'Not for human use; for animal treatment only' and shall bear a symbol depicting the head of a domestic animal.
4. The container of a medicine prepared for treatment of human ailments shall if the medicine contains industrial methylated spirit, indicated this fact on the label and be labeled with the words –

   "For External Use Only".

5. Substances specified in schedule X in bulk form shall bear a label wherein the symbol as specified in sub – rule (1) shall be given conspicuously in red letters.

Rule 102. Non-Sterile Surgical Ligature and Suture: Every container of, and wrapper enclosing surgical ligature of suture other than a ligature or suture offered or intended to be offered for sale as sterile, shall bear a label on which are printed or written in a conspicuous manner in indelible red ink the words " Non – sterile surgical ligature (suture) – not to be used for operations upon the human body unless efficiently sterilized".

1. Omitted
2. The name and address of the manufacturer shall be printed on the label of the container of a patent or proprietary medicine.
3. The true formula or list of the ingredient shall be printed or written in indelible ink on the outer label of every package containing patent or proprietary medicine.

Rule 104. Use of letters I.P., etc: the letters 'I.P. and recognized abbreviations of pharmacopoeias and official compendia of drug standards prescribed under these rules shall be entered on the label of the drug only for the purpose of indicating that the drug is in accordance with standards set out in the Indian Pharmacopoeia or in any such pharmacopoeia or official compendium of drug standards recognized under the rules.

Rule 104A. Prohibition against altering inscriptions on containers, labels or wrappers of drug: No person shall alter, obliterate or deface any inscription or mark made or recorded by the manufacturer on the container, label or wrapper of any drug:

[Provided that nothing in this rule shall apply to any alteration, any inscription or mark made on the container, label or wrapper of any drug at the instance or direction or with the permission of the licensing authority.]

Rule 105. Packing of drugs:

1. the pack sizes of drugs meant for retail sale shall be as prescribed in schedule P1 to these rules.
2. The pack sizes of drugs not covered by the schedule P1 shall be as given below:

   Unless specified otherwise in schedule P1,

   (i) The pack sizes for Tablets/ Capsules shall be -

   Where the number of Tablets (coated or uncoated)/Capsules (hard or soft gelatin) is less than 10, such packing shall be made by the integral number. or numbers above 10, the pack sizes of Tablets/Capsules shall contain multiples of five.

(ii) The pack sizes for liquid Oral preparations shall be 30ml. (pediatric only)

60ml. / 100ml./200ml. / 450 ml.

(iii) The pack sizes for pediatric Oral Drops shall be 5 ml./ 10 ml. / 15 ml.

(iv) The pack sizes for Eye/Nasal/ Ear drops shall be 3 ml. / 5 ml / 10 ml.

(v) The pack sizes for eye Ointment shall be 3 gm/5 gm / 1`0 gm.:

Provided that the provisions of the pack sizes covered under this rule shall not apply to –

1. Pack sizes of dosage forms not covered by the foregoing provisions of this rule
2. The imported formulations in finished form.
3. Preparations intended for veterinary use.
4. Preparations intended for Export.
5. Vitamins/Tonics/Cough Preparations/Antacids/Laxatives in Liquid Oral forms, Unit dose (including applicaps).
6. Pack sizes of dosages forms meant for retail sale to Hospitals, Registered Medical Practitioners, Nursing Homes.
7. Physician's Samples.
8. Pack sizes of Large Volume intravenous Fluids:

[Provided also that pack sizes of any of the new drug as and when approved by the licensing authority appointed under rule 21 and if not covered under this rule, shall be examined for the purpose of approval with specific justification by this said licensing authority.]

[Provide further that Oxytocin injection meant for sale shall be in single unit blister pack only.]

Rule 105A. Packing of drugs specified in schedule X: The drugs specified in schedule X shall be marketed in packing not exceeding –

(i) 100 unit doses in the case of tablets/ capsules:

(ii) 300 ml. in the case of oral liquid preparation;

(iii) and 5 ml. in the case of injections:

[Provided that nothing in this rule shall apply to packing meant for use of a hospital or a dispensary subject to the conditions that –

(i) such supplies are made by the manufactures of distributors direct to the hospital / dispensaries; and

(ii) hospital packs shall not be supplied to a retail dealer or to a Registered Medical Practitioner.]

Rule 106 Diseases which a drug may not purport to prevent or cure

1. No drug may purport or claim to prevent or cure or may convey to the intending user thereof any idea that it may prevent or cure, one or more of the diseases or ailments specified in schedule J.
2. No drug may purport or claim to procure or assist to procure, or may convey to the intending user thereof any idea that if may procure or assist to procure, miscarriage in women.

CHAPTER 16

Blow-Fill-Seal (BFS) in Pharmaceutical Packaging

Summary

16.1 Introduction

Pharmaceutical liquids can be filled and packaged in a number of different containers, including bottles, vials, ampoules made from glass or plastic or plastic bags. Whatever the choice of container, the role of pharmaceutical liquid packaging is to protect the contents from the outside environment during transportation and storage and to allow for safe and easy administration of the pharmaceutical liquid. Although production and handling of plastic containers is not without risk, a number of factors give valuable benefits to the producer and end-user. One of the most promising manufacturing technologies is certainly aseptic blow-fill-seal (BFS) technology. Blow-fill-seal processes create plastic vials, fill them with aseptic drugs, and seal them, all with minimal human interaction.

16.2 Blow-Fill-Seal (BFS) Technology

Blow-Fill-Seal technology refers to the manufacturing technique used to produce small, (0.1 m*l*) and large volume, (500 m*l* +) liquid filled containers. The basic concept of blow fill seal (BFS) is that a container is formed, filled, and sealed in a continuous process without human intervention, in a sterile enclosed area inside a machine. Thus this technology can be used to Aseptically Manufacture sterile pharmaceutical liquid dosage form.

16.3 Materials used for BFS

Generally the plastic containers are made up of polyethylene and polypropylene. Polypropylene is more commonly used to form containers which are further sterilized by autoclaving as Polypropylene has greater thermostability. The typical materials used in BFS processing are polyethylene (PE) or polypropylene (PP). Both are considered inert and give a good balance of properties, enabling easy forming, opening and handling of the finished container.

16.4 Plastic vs. Glass Containers

Glass, although a standard in the aseptic pharmaceutical liquids industry, is not without its limitations. There is the safety issue - glass vials are subject to breakage, both in transit and while being administered. Handling glass containers always involves a certain amount of risk of lacerations and glass splinters. Glass ampoules, for example, generate a fine array of small glass particles during opening. It has become more cost effective to use plastic containers for aseptic liquids, which effectively costs manufacturers one-third of the cost of glass. Plastic is less expensive to ship because the containers are lighter. For small-volume parenterals, the use of plastic is inevitable, and increasingly being considered for these reasons.

16.5 Microbial and Particulate Integrity BFS System

The BFS critical processing zone is continually supplied with HEPA-filtered air by an air shower device (shroud). The BFS critical zone is the area where the containers are exposed during filling. Air in the critical zone meets Class 100 (ISO 5) microbiological standards during operations. Sterility of BFS polymeric containers, materials and processes is validated by verifying that time and temperature conditions of the extrusion, filling and sealing processes are effective against endotoxins and spores. Challenge studies have been conducted on the sterility levels of advanced BFS technology, which demonstrate a uniform capability of achieving contamination rates not exceeding 0.001 percent throughout the entire process. Even higher sterility assurance levels, approaching 0.000001 percent, have been achieved using high levels of airborne microbiological challenge particles.

16.6 Conclusion

BFS technology has established itself as an efficient, advanced aseptic processing technology for pharmaceutical liquids and semisolids. The technology is best suited for high production volumes and will provide a reliable means of producing aseptic products. BFS technology for pharmaceutical liquids not only provides the manufacturer with a safe and efficient production method for aseptic products, but it will also create possibilities for innovative design, branding and product life-cycle management.

16.1 Introduction

Pharmaceutical liquids can be filled and packaged in a number of different containers, including bottles, vials, ampoules made from glass or plastic or plastic bags. Whatever the choice of container, the role of pharmaceutical liquid packaging is to protect the contents from the outside environment during transportation and storage and to allow for safe and easy administration of the pharmaceutical liquid. Depending on the active ingredient and other components of the formulation, it is not unusual for degradation of the product to occur, either during processing, long-term storage or both. Many of the new drug compounds made via biotechnological routes do not, for example, withstand steam sterilization or irradiation and so are best treated aseptically. Traditionally, glass has been the dominating packaging material for sterile pharmaceutical liquids; however, aseptic processing, involving filling open glass bottles or vials, sets high demands on the manufacturer to maintain aseptic conditions at all time, especially for large-volume containers. Handling glass containers always involves a certain amount of risk of lacerations and glass splinters. It has also been shown that glass ampoules generate a fine array of small glass particles during opening. The sub-atmospheric pressure inside the ampoule tends to draw such particles inside the ampoule on opening, particles that, at best, are sterile, or, in the worst case, are contaminated by the person handling the ampoule. Although production and handling of plastic containers is not without risk, a number of factors give valuable benefits to the producer and end-user. One of the most promising manufacturing technologies is certainly aseptic blow-fill-seal (BFS) technology.

16.2 Blow-Fill-Seal (BFS) Technology

Blow-Fill-Seal technology refers to the manufacturing technique used to produce small, (0.1 m*l*) and large volume, (500 m*l* +) liquid filled containers. Blow-Fill-Seal technology was originally developed in Europe in the 1930s and was introduced in the United States in the 1960s, but over the last 20 years it has become more prevalent within the pharmaceutical industry, and it is now widely considered to be the superior form of aseptic processing by various medicine regulatory agencies including the U.S. Food and Drug Administration (FDA) in the packaging of pharmaceutical and healthcare products. The first BFS machines were constructed in the 1960s, although machines for pharmaceutical applications were first introduced in the 1970s. The BFS machine is constructed from such vital components as:

- a polymer granule storage and feeding system;

- a rotating screw extruder with parison head;
- a sterile air filling chamber;
- mould halves to form and close the container; and
- downstream equipment including, for example, leak detecting systems.

The basic concept of blow fill seal (BFS) is that a container is formed, filled, and sealed in a continuous process without human intervention, in a sterile enclosed area inside a machine. Thus this technology can be used to Aseptically Manufacture sterile pharmaceutical liquid dosage forms.

The process is multi-stepped, firstly pharmaceutical-grade plastic resin is vertically heat extruded through a circular throat to form a hanging tube called the **Parison**. This extruded tube is then enclosed within a two-part mould, and the tube is cut above the mould. The mould is transferred to the filling zone, or sterile filling space where filling needles **mandrels** are lowered and used to inflate the plastic to form the container within the mould. Following the formation of the container, the mandrel is used to fill the container with liquid, following filling the mandrels are retracted and a secondary top mould seals the container. All actions take place inside a sterile shrouded chamber inside the machine.

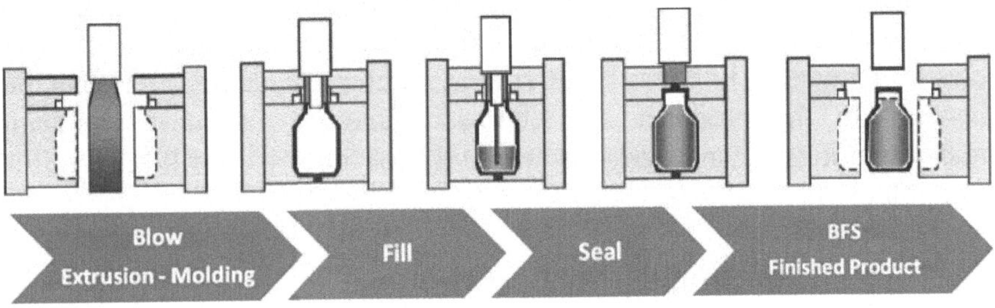

Fig.16.1 Various steps in BFS system.

The product is then discharged to a non-sterile area for labeling, packaging and distribution. Thus, the forming, filling and sealing steps have been achieved in one unit operation–the cycle being completed within 10–12 seconds. The complete aseptic system is typically cleaned and steam sterilized in place, and no production personnel are present in the filling room during normal operation. If the machine is run in an improper (non-ideal) fashion, the risk of leaking containers increases. However, any signs of leakage can be detected effectively via 100% in-process leak detecting systems using a high voltage.

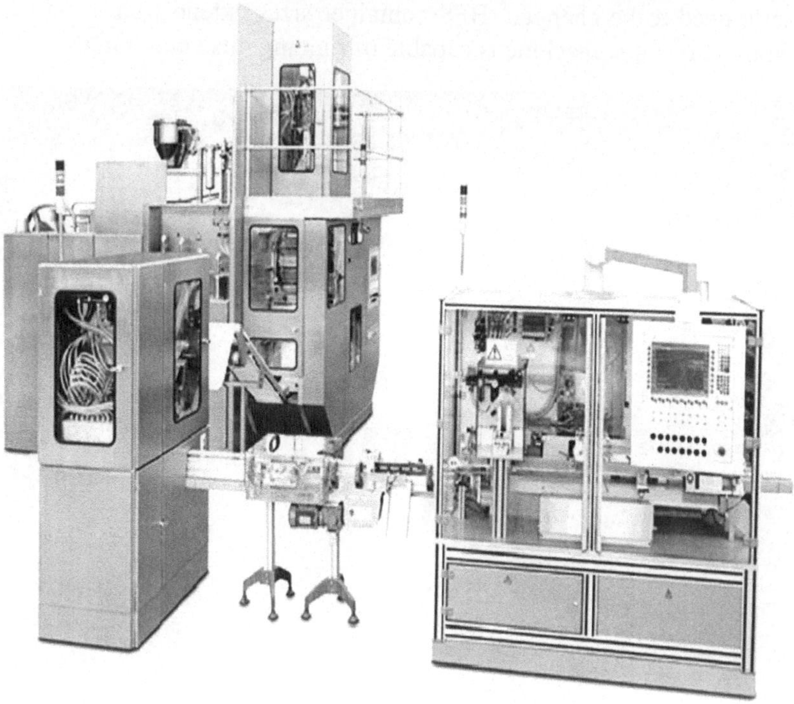

Fig. 16.2 Liquid container filling and sealing machine with BFS system

Blow-fill-seal technology reduces personnel intervention making it a more robust method for the aseptic preparation of sterile pharmaceuticals. Highly versatile, BFS is compatible with many products, container sizes, and packaging formats, including unit-dose containers. Virtually any viscous or suspension fluid can be accommodated, and liquid-handling systems can be configured to blend multi-ingredient formulations. BFS is used for the filling of vials for parenteral preparations and infusions, ophthalmic and inhalation products. BFS container options include vials, ampoules, bottles, plunger-less syringes, and custom containers. Designs often are squeezable and may be equipped with bellows to aid dispensing. Containers also can incorporate a number of built-in features such as twist-off opening; dropper, stopper, piston, or cannula inserts; cutting-ring or self-piercing threaded closures; and Euro-Caps for infusion bottles. Changeover time varies. If a BFS machine is used to make a family of containers, changing mold inserts to switch between 2-, 3- or 5-m*l* sizes, for example, can take two to four hours. Full mold changes will take longer, probably about eight hours. However, many pharmaceutical packaging lines minimize changeover downtime by dedicating machines to a single style and size of container or by running a narrow size range so only mold inserts need to be changed. BFS container sizes extend from 0.1 to 10 L, although no single machine is capable of running this entire range.

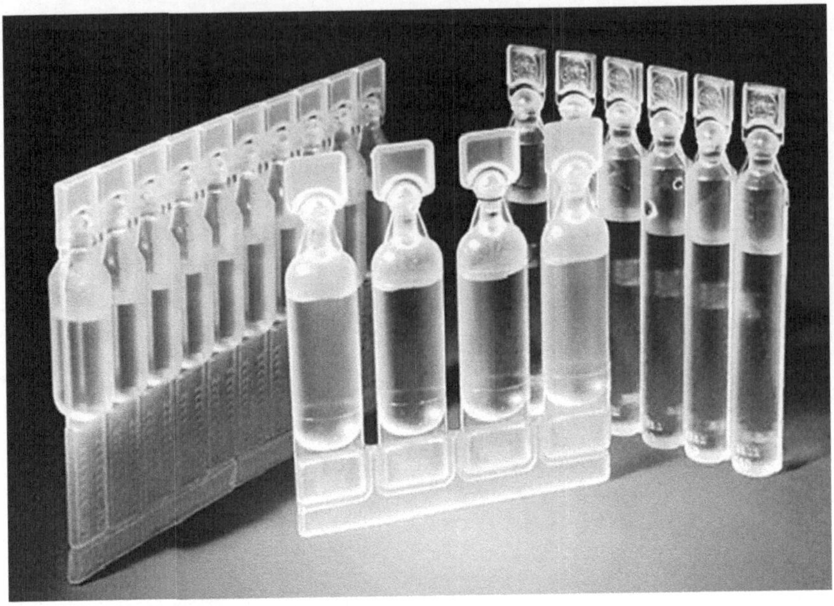

Fig. 16.3 Package filled and sealed with BFS system.

16.3 Materials used for BFS

Generally the plastic containers are made up of polyethylene and polypropylene. Polypropylene is more commonly used to form containers which are further sterilized by autoclaving as Polypropylene has greater thermostability. The typical materials used in BFS processing are polyethylene (PE) or polypropylene (PP). Both are considered inert and give a good balance of properties, enabling easy forming, opening and handling of the finished container. Because there is some amount of crystallinity, the materials are somewhat cloudy, although transparent enough to allow for visual inspection of particulates. As with all polymer materials, care must be taken during development to ensure that no ab/adsorption of substances from the formulation is taking place, especially with lipophilic drugs. Migration of unwanted components from the polymer materials is normally not an issue if well- known medical-grade polymers is used. Another point to consider when working with polymer material, is the risk of oxidative degradation of the formulation. This can be resolved by using foil envelopes or possibly co-extrusion including barrier resins. Since PE and PP consist of carbon and hydrogen only, combustion of the materials generates water and carbon dioxide, rendering them energy-recoverable. Performing a life-cycle assessment (LCA) is a common method of evaluating how various products affect the environment. Several studies performed in recent years provide evidence that less harmful effects on the environment are associated with the production, transportation and disposal of polymer containers than with glass containers. A potential downside when working with plastic materials is that suppliers tend to withdraw certain products from the market as polymer technology advances or companies merge. Such withdrawals might cause costly re-registrations of equivalent plastic materials for the pharmaceutical manufacturer. To maximize security of supply, it might therefore be advisable to use and register two equivalent polymer materials, should one of them be removed from the market.

16.4 Plastic vs. Glass Containers

Injectables, ophthalmics, biologicals and vaccines are produced in a number of different types of containers, including bottles, vials and ampoules that are made from glass and plastic. Protecting the contents of these aseptic liquid drugs through filling, packaging and transportation, and allowing for safe and easy administration are critical objectives in the aseptic process. The industry is infused with a strong quality control emphasis. Raw materials, and in-process and finished products are continually checked for approval and rejection. The packaging needs for pharmaceutical liquids are quite demanding. It is not unusual for degradation of the product to occur during processing or while in transit. The physical properties of liquids can be altered with inadequate packaging components. For aseptic filling, the package must be produced, stored, filled and sealed under conditions that preserve sterility. Likewise, the appearance of particulates in sterile solutions is equally undesirable. Glass, although a standard in the aseptic pharmaceutical liquids industry, is not without its limitations. There is the safety issue - glass vials are subject to breakage, both in transit and while being administered. Handling glass containers always involves a certain amount of risk of lacerations and glass splinters. Glass ampoules, for example, generate a fine array of small glass particles during opening. Manufacturers using glass containers are also subjected to design limitations when the designs become somewhat complex. With glass containers, as design complexity increases so does the cost. Once glass containers are produced, they need to be transported to the aseptic facility. Glass is typically transported in cardboard boxes that can contain mold spores, such as Penicillin sp. and Aspergillus sp., as well as bacteria like Bacillus sp. Paper used in the shipping of glass, can also contain mold spores. The rubber closures used on the glass containers can have mold contamination. Domestic drug companies have been slow to change to plastic, primarily due to the existing installed base of glass production of small-volume parenteral drugs in the United States. However, the same is not the case with new drugs that are coming onto the market. These are more frequently being looked at, and submitted for FDA approval, in plastic containers produced by advanced BFS aseptic processing. Supporting this move is that the BFS processing resins, polyethylene and polypropylene, are generally considered inert by the FDA. Many of the blow molding resins used in BFS processing have received international acceptance as suitable for food and drug applications, and many of the drug products produced outside of the United States can be found packaged with these resins. With the continued refinement of BSF technology, its acknowledgment by the FDA as a preferred technology for aseptic processing, and its growing acceptance by drug companies, the migration from glass to plastic containers used for aseptic pharmaceutical liquids is growing rapidly. It has become more cost effective to use plastic containers for aseptic liquids, which effectively costs manufacturers one-third of the cost of glass. Plastic is less expensive to ship because the containers are lighter. For small-volume parenterals, the use of plastic is inevitable, and increasingly being considered for these reasons. Although many BFS systems make available only a limited number of container choices within each container category, some BFS machines do allow for broad versatility in container design. Advanced BFS machines can design virtually any container mold through the use of sophisticated CAD/CAM technology and 3-D modeling. These design systems, when interfaced with the latest in CNC and EDM machinery, ensure fabrication of key components to precise tolerances. BFS machine designs also allow for mounting of separate sterile items (inserts) within the BFS container, and in-mold coding and engraving, which provide further opportunities for innovative design over that of glass products.

16.5 Microbial and Particulate Integrity BFS System

Sterility of BFS polymeric containers, materials and processes is validated by verifying that time and temperature conditions of the extrusion, filling and sealing processes are effective against endotoxins and spores. Challenge studies have been conducted on the sterility levels of advanced BFS technology, which demonstrate a uniform capability of achieving contamination rates not exceeding 0.001 percent throughout the entire process. Even higher sterility assurance levels, approaching 0.000001 percent, have been achieved using high levels of airborne microbiological challenge particles. Endotoxins are a potential pyrogenic contaminant, essentially dead bacterial cellular matter. They can lead to serious reactions in patients, particularly with those receiving injections, ranging from fever to death. A critical aspect of BFS technology is its pyrogen-free molding of containers and ampoules. Extensive experiments confirming the efficacy of the BFS extrusion process have been performed using high levels of spores and endotoxin-contaminated polymer granules. The typical BFS extruders have demonstrated spore contamination rates of 0.000001 percent, and 0.00001 percent for endotoxins. Control of air quality is critical for sterile drug product manufacture. BFS equipment design typically employs the use of specialized measures to reduce microbial contamination and particle levels that can contaminate the exposed product. The BFS process inherently produces a very low level of particulate matter and much of potential BFS microbial contamination (viable) in the air is mitigated by the absence of manual intervention in its critical areas. Non-viable particles generated during the plastic extrusion, cutting, and sealing processes are controlled. Provisions for carefully controlled airflow protect the product by forcing created particles outward while preventing any inflow from the adjacent environment. These "zones of protection" can also incorporate designs that separate them from the surrounding environment, providing additional product protection. The BFS critical processing zone is continually supplied with HEPA-filtered air by an air shower device (shroud). The BFS critical zone is the area where the containers are exposed during filling. Air in the critical zone meets Class 100 (ISO 5) microbiological standards during operations. The critical zone is continuously monitored to ensure a positive differential pressure maintained between the shroud and the adjacent clean room.

16.6 Conclusion

BFS technology has established itself as an efficient, advanced aseptic processing technology for pharmaceutical liquids and semisolids. The technology is best suited for high production volumes and will provide a reliable means of producing aseptic products. The increasing focus on branding, needle stick prevention, safety and convenience in handling of pharmaceutical products will create incentives for the more widespread use of BFS technology. New, improved machine models are constantly being produced, allowing higher filling speeds, better quality and/or improved aseptic control. New, softer and more transparent polyolefinic materials are also under evaluation. The increasing environmental awareness also reflects on the use of pharmaceutical packaging and impact, for example branding. LCA studies can be used to evaluate the environmental impact of products, and BFS ampoules generally show good environmental profiles. BFS technology for pharmaceutical liquids not only provides the manufacturer with a safe and efficient production method for aseptic products, but it will also create possibilities for innovative design, branding and product life-cycle management. As long as the compatibility between formulation and container material is proven, BFS technology will provide many valuable advantages to both producer and end-user.

APPENDIX

Some Packaging Machines and Specifications

Blister Packaging Machine

Blister packaging machines are used by the pharmaceutical industry to pack capsule and tablets.

The packing process initiates with the capsules or tablets being loaded into a hopper and then into a feeder, which in turn can either be linear feeder or a brush box feeder depending on the shape of the product and also the material to be used. Heated rollers in the machine use different plastic films to form blisters that are subsequently filled with tablets. Some machines make use of Aluminium foil to form blisters. Machines are manufactured according to the size, depth and material used to form blisters. After tablets have been inserted in the formed blisters, cavities are heat sealed using foil.

Types of Blister Packaging Machine

- **Blister Sealing Machines** are used to seal filled plastic blister to a piece of coated carton board, by heat. The blister is kept in the sealing die which has the same shape as that of the cavity. The cavities are then filled and sealed by a foil using heat.
- **Blister Trimming Machines** are used for cutting / trimming the sealed blister or formed blister packs in individual pieces or bunch.

High Speed Automatic Bottle Filling and Cap Sealing Machine

This is a volumetric principle-filling machine. This unit is fully automatic unit and can be coupled with an unscramble to feed the bottles in feed conveyor of this machine. This machine achieves the complete operation of filling the bottles automatically and accurately.

The complete filling sequence for different fill sizes is programmable for the speed with the movement of a switch, achieves complete synchronization automatically. The fill range of this machine can be adjusted from 30 ml to 1000 ml with the change of syringes. Higher fill volumes can be achieved by taking multiple strokes of the syringes. The number of filling head can be either 4, 6 or 8 depending upon the output required from this unit. Usually we get an output of 80-160 bottles per minute. This also depends on the fill volume of the bottle and number of filling heads.

The filling machine is further attached to a manual capping machine (pedal type) which is to be operated by an operator who would carry the bottle from the filling section, place the cap on the bottles neck and seal it with the help of the rotating head. This operation is also done automatically by connecting an online automatic bottle capping machine.

The sealed bottles are then carried towards the inspection table by means of a slat conveyor, where the bottles are visually inspected by the operator for any particles in the filled liquid. This process is simultaneously carried out in a white and black background and thereafter the bottles are further transferred for next operation, i.e. labelling / packaging.

This machine can be adapted to a wide variety of bottle shapes and liquids properties and can be used in various industries such as Pharmaceuticals, Cosmetics, Agrochemicals, Consumer goods, Chemicals, etc. This machine can be used for any type of liquids.

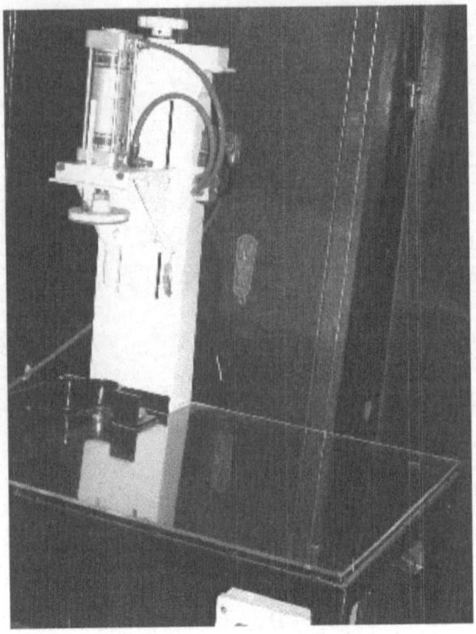

Twin Head Volumetric Filling Machine

The compact designed unit consists of two stainless steel syringes with non return valve and HDPE piston ring for accurate filling.

Output: 4000 to 10,000 bottles per shift.

Suitable for filling all kinds of liquids in Pharmaceutical and Cosmetic industries

Automatic Labelling / Gumming / Stikering Machine

Automatically Labels on Round Containers of Glass and Plastic, with Wrap Round Labels and also Flag Type in the Vertical Position by using Cold Glue.

This Machine is rugged, versatile and engineered for reliability and enhances operational efficiency.

In addition, it has several exclusive features which offer additional advantages.

Hundreds of units installed in the Pharmaceuticals, Cosmetics, Food Plants and also Liquor Industries are boosting profits by giving the products a distinctive look.

Syringe Filling Machine

Syringe filling machines are high precision and reliable machines used to fill syringe, cartridges and other related containers. These fillers are manufactured to operate at both low and high speeds for both inline as well as nested filling process. High technology are equipped with features such as barrier isolators, in-process check weighing and safety sensors. Filling is normally done with the help of rotary piston pumps. The machines format spectrum can range from 0.2 to 29 ml.

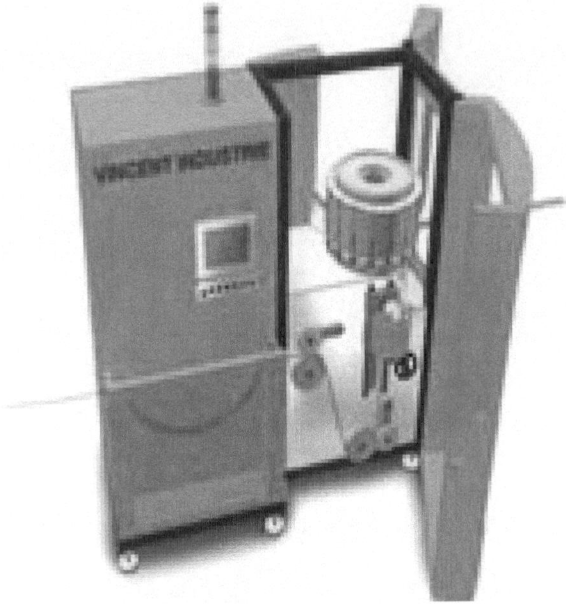

Semi Automatic Syringe Filling Machine: These machines require manual operators for loading the syringes into the machine which are then filled and capped automatically. Applications include oral dosage syringe and dental gels.

Fully Automatic Syringe Filling Machine: These high speed and compact machines automatically fill and cap the syringes and require minimum operator interference and are used for saline flush syringe, dental gels and oral dose syringe.

Types of Syringe Filling Machines

Inline Filling: All types of syringe including glass, plastic, tips (with or without flanges) can be filled. Depending on the kind and features, theses machines can fill anywhere between 300 to 600 syringes in a minute.

Nested Filling: These syringe filling machines are highly sophisticated and fully automatic and can be used to fill and seal pre- sterilized nested syringes.

Ampoule Filling Machine

Ampoule filling machines are extensively used by the pharmaceutical industry. These high precision machines completely encase the product in the inert glass and do not have a rubber stopper or any other material in direct contact with the drug. These flexible machines can be used to fill ampoules as well as vials and the speed varies according to the technology. Certain ampoule coding machines apply ring banding to the completed ampoule tips for accurate identification. Filling range of these machines is normally between 1ml to 20 ml.

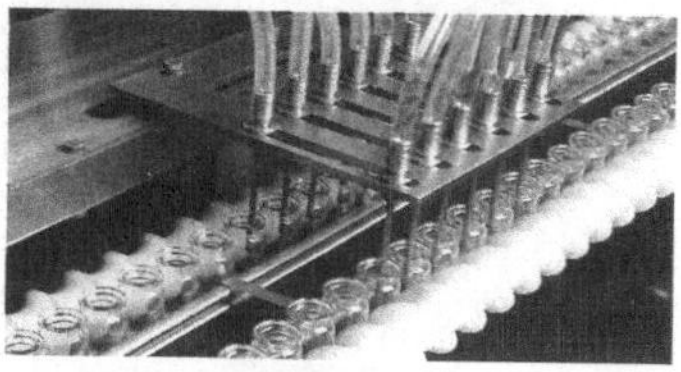

Features of Ampoule Filling Machines

Most ampoule fillers are characterized with fast changeover time to accommodate a variety of ampoules in terms of shapes and size. These machines require minimal maintenance and are very easy to clean. Another important feature is the installation of speed adjusting equipment and 'no ampoule no fill' capability to ensure that unfilled ampoules are not packed. Check weight mechanism of the machine helps to maintain consistency in each batch. Sealing is done either by laser sealing system or conventional gas flame.

Application of Ampoule Filling Machine

- Pharmaceutical
- Neutraceutical

Cartoners

Cartoners are specialized pharmaceutical packaging machines that can accommodate paperboard and corrugated material and are used for a number of applications. These machines are suited for bag-in-box application and are easy and simple to operate. Semi automatic and compact cartoning equipment include features like hand load, automatic bottle loading, volumetric, auger and robotic filling. These machines are capable of manufacture different sized/styles cartons. Cartoners have an output ranging 30 to 300 cartons per minute (CPM) depending on whether the machine is vertical loading, intermittent cartoning or a continuous motion model. These machines can handle blister strips and other pharmaceutical packing.

Types of Cartoners

- **Horizontal Cartoning**: This machine erects a folded carton, filling it with a product horizontally through an open end and then closing it by tucking the end flaps of the carton or by applying adhesive. These machines are used for packaging confectionery, food stuff, medicines, cosmetics etc.
- **Vertical Cartoning**: This machine erects a folded carton, filling it with a product vertically through an open end and then closing it by tucking the end flaps of the carton or by applying adhesive. These machines are used for packaging confectionery, food stuff, medicines, cosmetics etc.
- **Wrap-around Cartoning**: This machine wraps a carton board around a product and secures it by applying adhesive or by engaging pre cut tabs. Mostly used for medicines and foodstuff.

Shrink Wrap Machines

Shrink wrap machines make use of plastic to effectively seal a container or product by tightly surrounding it. Vacuum shrink wrapping can prevent bacteria form entering the product and also protect them from wear and tear during transportation. The process involves tightening a thermoplastic film around the product and shrinking it by heat. It also helps to add strength to the stack, making it dust, water and tamper proof. These machines are used for pharmaceutical packaging. Certain shrink packaging machines perform the dual function of shrinking and sealing.

There are various types of shrink wrap machines used according to the product or material being wrapped.

- L-Bar sealers are used to pack products with the help of electricity rather than heat.
- Heat guns are used to wrap blister packs.
- Vacuum sealers are used to shrink wrap pharmaceutical, neutraceutical and medicines.

Automatic Bottle / Container Capping Machine

It is designed to seal metallic caps / Plastic screw caps of various sizes on the bottles fed on continuous line. It is fully automatic and is of robust construction, complete with a rotary cap feeding hopper and a built-in slat conveyor for the incoming and outgoing bottles.

An in feed worm and star wheel of appropriate size, transfer the incoming bottles with caps and feed it under the sealing head. The synchronized movement of the bottles and the sealing mechanism provides perfect sealing. The sealed bottles are then further transported by another star wheel and outgoing slat conveyor of the machine.

Name and Addresses of some manufacturers.

S.NO.	Packing Machines	Name and Address
1	Injectable Drugs	West Pharmaceutic Servives, Singapore Alagu.Subbu@west pharma.com
2	Labelling Machine	Ambica Pharma machines Pvt. Ltd. Plot-1, Phase II GIDC Valva, Ahmedabad www.ambica.co.in
3	Preprogrammed Electronic Cosmetic Package/Bundlers-Overwrapper	Kavai Pharma Machineries 81 J.K. Industrial Estate Mahakali Caves Road Andheri CE, Mumbai 13 Sales@kovaipharma.com

Contd...

S.NO.	Packing Machines	Name and Address
4	Dropper Bottle with Philpher Proof Twin Cap	Steri pack 101, Bank of Maharastra Lane. 7, Bungalows Andheri (w) Mumbai 61 www.steripack.com
5	Ampule and vials coding on label application machine	Condot Systems Pvt Ltd. www.condotsystems.com
6	Aluminium foil	Rolls Pack Pvt Ltd/Unipal Pvt Ltd Unit No-1 Minorva Industrial estate mulana(w), Mumbai www.rollspace.com
7	Processed Wood Pallets	Shree Umiga Timber Industries Plot. 152-153, Industrial Estate GIDC Gundlav, Valsad umigaind@gmail.com
8	Liquid Filling Machine/ Sealing Machine	National Pharma machinery 11 Varahi Estate, Nr. Railway Crossing, Gota Road Gota Ahmedabad 81 www.nationalindia.co.in
9.	Carton Sealing Machine/ Strapping Machines Coding Machines	Packomark Technologies Pvt Ltd. C-266, Sector 63, Noida 81 (UP) Info@packomark.com
10	Caps, Septals Crimpers and Decappers	La-Pha-Pack Raj Analytical Plot 601, Ind. Focal point phase IX Distt. Mohali – 160062 www.rajanalytical.com
11	Collating and overrapping Machine/Capping Machine	Autopack Machine Pvt Ltd 101-C, Poonam Chambers Worli Mumbai-18 www.autopackmachines.com
12	Glass/Plastic bootles Tubes/Vials Closures	Penn bottles and Supply Co. S. Zhaveri Pharmaceut Pvt Ltd. 109, Shiv Smriti Chambers 49 Dr. A.B. Road, Mumbai-18 www.szhaveri.com

Contd...

S.NO.	Packing Machines	Name and Address
13	Tube Packaging	Subnil packaging machineries www.subnil.com
14	Carton/Bundlers	Vectacraft www.vectacraft.com
15	Cartoning Machine	Harikrishna machintech Pvt Ltd. Plot-C-1, 4310, Phase IV GIDC Vatva, Ahmedabad 45 www.harikrishna.com
16	Ampule filling and Sealing Machine	Snowbell Machine Pvt Ltd. Unit-2,3,4, Atul Estate Sativali, Vasai(E) Thane-08 www.snowbellmachines.com
17	Blister Packing Machine	Teknopak Filtra teknopak cleanroom system Pvt Ltd. Mumbai Teknopak@vsnl.com www.teknopkpnline.com
18	Eye, Nasal drop filling System	Techline Industries A-105 Bharat Industrial estate Bhandup (w) Mumbai-78 http://techlineindustries.com
19	PVDC Coating PVC Film	Caprihans India Ltd Block-D, Shivsagar Estate Worli Mumbai-18 www.caprihansindia.com
20	Prefill Syringes	Sewa Medicals Ltd. 10-11 Proped-Chambers Annex Dr. D.N. Road- Mumbai-1 www.sewamedicals.com

Index

A

B

D

E

F

G

H

Q

R

S

T

9 789385 433269

Printed at Repro India Ltd.